An Introduction to Unconstrained Optimisation

A Computer Illustrated Text

AN INTRODUCTION TO UNCONSTRAINED OPTIMISATION

J J McKeown, D Meegan and D Sprevak

Department of Engineering Mathematics,
The Queen's University of Belfast

CRC Press

Taylor & Francis Group

Boca Raton   London   New York

CRC Press is an imprint of the
Taylor & Francis Group, an **informa** business

First published 1990 by IOP Publishing Ltd

Published 2020 by CRC Press
Taylor & Francis Group
6000 Broken Sound Parkway NW, Suite 300
Boca Raton, FL 33487-2742

© 1990 by Taylor & Francis Group, LLC
CRC Press is an imprint of Taylor & Francis Group, an Informa business

No claim to original U.S. Government works

ISBN 13: 978-0-7503-0025-4 (pbk)
ISBN 13: 978-1-138-40412-0 (hbk)

Visit the Taylor & Francis Web site at
http://www.taylorandfrancis.com

and the CRC Press Web site at
http://www.crcpress.com

British Library Cataloguing in Publication Data

McKeown, J.J.
    An introduction to unconstrained optimisation
    1. Unconstrained optimisation. Numerical methods
    I. Title   II. Meegan, D.   III. Sprevak, D.   IV. Series
    515

ISBN 0-7503-0025-6
ISBN 0-7503-0026-4 IBM disc
ISBN 0-7503-0027-2 BBC 40/80 track disc
ISBN 0-7503-0028-0 text and disc
ISBN 0-7503-0029-9 Network pack

Library of Congress Cataloging-in-Publication Data are available

Contents

Preface

This book is an addition to a series which uses microcomputers as an aid to the understanding of mathematical techniques. Numerical Optimisation is a topic well suited to this approach, most of the algorithms used being based upon a concept of "hill-climbing" which, in two dimensions, is easily illustrated dynamically on the computer screen. It has thus proved possible both to describe the key methods and to explain the motivation behind the developments which have taken place in recent years.

The treatment is appropriate to a first course in numerical analysis. In this context it may be used as a self-teaching text by those with some basic knowledge of differential calculus and vector algebra. Alternatively, it can be used selectively under the guidance of the teacher or lecturer.

In addition, with numerical optimisation becoming more and more widely accepted as a powerful approach in many application areas, the book is intended to be useful to anyone wishing to become rapidly familiar with the key ideas and methods before applying them in his own field. To this end, and to preserve the compact format which is a feature of this series, rigorous proofs and comprehensive lists of methods have been kept to a minimum in favour of a clear line of development.

The programs have been integrated into the text in a number of ways. Firstly, examples and exercises using the computer have been provided liberally throughout the text, and the illustrations on the page are those which the reader can reproduce for himself on the screen. Secondly, by making the interface flexible and modular, we have provided a means by which the reader can explore the algorithms further and even try alternative approaches for himself. He is therefore in a position to verify the points made in the text rather than accepting them at face value.

We hope that this small textbook will open up the field of numerical optimisation to a wider audience.

J J McKeown, D Meegan and D Sprevak
Belfast 1989

> Chapter 1

> Getting Started

> 1.1   Installing the software

The software associated with this book is supplied on a floppy disc. This disc is write protected and should be used to make a working copy which **must not be write protected** and will be used to run the programs. The original copy should then be kept in a safe place.

The information required to make a working copy on your system is contained in a README file on the original disc. View this file and proceed as directed.

> 1.2   Starting the software

To start the package, insert the working copy into the disc drive and activate the menu program following the procedure given in the README file.

> 1.3   Activating programs

The first program in the menu is used to prepare functions and make them available to other programs. It is therefore an appropriate choice to illustrate the use of the package.

Select PREPARE by placing the selection bar on the first item in the menu and pressing the **RETURN** key. The screen will clear briefly while PREPARE is being loaded from disc. On completion of this process, a new menu, we will call it the main menu, will be displayed on

1

the first line of the screen. The second line will display a message to
indicate that a function has not been prepared.

*The first line is always used to request information while the
second line is always used to display information or results.*

The rectangle in the bottom left quadrant of the screen is used to
display contour plots; this facility will be discussed in Chapter 2.

As a function has yet to be prepared, the only valid options at this
stage are **E(dit**, **L(oad**, and **Q(uit**. The **Q(uit** option is used to return
to the Menu program, **L(oad** to load a function from a disc, and **E(dit**
is used to manipulate a function. When a function is available, it can
be saved to a disc file with the **S(ave** option or displayed as a contour
plot with the **D(isplay** option.

Start with the **E(dit** option. An option is selected by typing the
letter that appears before the bracket, so press the **E** key to select edit. A
new menu appears. This menu contains more options than will fit on the
40 column line and you can view the other options by pressing the **LEFT
ARROW** key or the **RIGHT ARROW** key to scroll through the
menu. **E(dit** is used to prepare a function. The preparation of a function
may consist of defining the function and its derivatives, and setting limits
on the dependant variables. However, defaults have already been set for
the limits and we do not need to define the derivatives at this stage.

In general, to enter a function the **F(X1,X2)** option is used.

However a different procedure for entering the function is required
when the function is formed by sum of squares and the least squares
method of Chapter 5 is to be used.

We discuss first the **F(X1,X2)** option. Let us enter the function:

$$F(x_1, x_2) = -x_2^2 + x_1^2 + 2x_1 + x_2 + (x_1 - x_2^2)^2$$

by activating the **F(X1,X2)** option; the request line changes to:

F(X1,X2)=

which is an invitation for you to enter the function. The function is
entered using the conventions of BBC Basic and, while you do not need
to be familiar with it to use the package, you must know how to enter
an expression in BBC Basic.[1] If you do not have this information you

---

[1] A peculiarity of BBC Basic is that the unary minus operator has a higher priority
than the exponentiation operator (^). So the expression –2 ^ 2+2 gives the result
6. If the exponentiation is to be performed first, the expression must be entered as
either –(2 ^ 2)+2 or 2–2 ^ 2, both of which give the expected value of –2.

should consult the relevent Acorn documentation. Enter the function by typing:

$$-(X2^{\wedge}2)+X1^{\wedge}2+2*X1+X2+(X1-X2^{\wedge}2)^{\wedge}2$$

Use the **DELETE** key to correct mistakes and the **RETURN** key to enter the function. If typing mistakes are found later, they can be easily corrected by selecting the **F(X1,X2)** option again. In addition to the prompt

F(X1,X2)=

being shown on the request line, the current setting is displayed on the information line.

The arrow keys can be used to move an edit cursor to any location on the information line and the **COPY** key may be used to copy information to the prompt line.

Having entered the function, you should return to the main menu by pressing the **RETURN** key at the menu prompt. A grid of dots will be displayed while the program scans the function over the limits set for x_1 and x_2. On completion of this process, it will have obtained a rough estimate for the range of function values covered by the region and will display this information as a table.

The option **S(quared function** is used to enter a function formed by a sum of squares. When activating this option you will be asked for the number of squared terms in the function. The screen also shows the default value for the option and the range of its allowed values. To input the function select now the option **F(X1,X2)** and enter the different squared terms in order.

From the main menu you can now select the **D(isplay** option to have the surface displayed as a contour plot. While the plot is being generated, the information line will show that the function is being calculated and stored in a disc file. Once it is stored on the disc, the contour plot can be displayed more quickly.

The number of contour bands which are displayed is fixed; so to show details of a surface in a region with a large range of contour values it may be necessary to make a plot with a reduced range of contour values. The option **H(igh F** in the **Edit** menu can be used to specify the highest value of the contour to be displayed. When **H(igh F** is activated the program calculates the function over a crude grid of points, after which it displays the average height of the contours and their range of values over the grid. The user can enter any value within the shown range, or simply by pressing the **RETURN** key select the average calculated height on the grid. If the option **H(igh F** is not activated the program uses the whole range of calculated values.

> Chapter 2

> Searching for an optimum

> 2.1 Introduction

Optimisation deals with the study of methods for solving problems of the type:

> Find the value of **x** such that some quantity F, which depends on **x**, is a minimum.

This is an equivalent problem to that of maximising $-F$; so we will refer only to the minimisation problem.

The quantity **x**, the independent variable, might be a single variable (a scalar) or a vector with n components, that is:

$$\mathbf{x} = \begin{bmatrix} x_1 \\ \vdots \\ x_n \end{bmatrix}$$

In the case where **x** has n components we say that F is a real valued function of n variables. For example:

$$F(x_1, x_2) = 1 - 2x_1 + 2x_1^2 + x_2^2 - 2x_1 x_2$$

is a function of two variables as it depends on the two-dimensional vector

$$\mathbf{x} = \begin{bmatrix} x_1 \\ x_2 \end{bmatrix}$$

5

and can be written:

$$F(\mathbf{x}) = 1 + \mathbf{x}^t \begin{bmatrix} -2 \\ 0 \end{bmatrix} + \mathbf{x}^t \begin{bmatrix} 2 & -1 \\ -1 & 1 \end{bmatrix} \mathbf{x}$$

where $\mathbf{x}^t$ is the transpose of $\mathbf{x}$.

When F and $\mathbf{x}$ take real values one can usually represent F by a graph, and indeed we will assume that there is a single value of F for each value of $\mathbf{x}$. When $\mathbf{x}$ is a scalar the graph consists of a curve; when $\mathbf{x}$ has two components F would be represented by a surface. For more than two dimensions it is difficult and not usually helpful to try to visualise the form of the function.

Recall that we wish to find a value of $\mathbf{x}$ which minimises F. In fact, we can distinguish two kinds of minimum. The first is simply the value of $\mathbf{x}$ for which F takes its lowest value, this is the 'global minimum' . The second kind, which we call a 'local minimum', and denote by $\mathbf{x}^*$, has the property that $F(\mathbf{x}^*)$ takes a smaller value than any $F(\mathbf{x})$ in the neighbourhood of $\mathbf{x}^*$. That is, in mathematical notation:

$\mathbf{x}^*$ *is a local minimum,*

if there is a positive number, ϵ, such that for all

$$0 < ||\mathbf{x} - \mathbf{x}^*|| < \epsilon$$

$$F(\mathbf{x}) > F(\mathbf{x}^*).$$

On the other hand:

A point $\mathbf{x}^*$ is the location of the global minimum of F if for all the allowed values of $\mathbf{x}$, $F(\mathbf{x}^*) < F(\mathbf{x})$.

The problem of global optimisation will be discussed in Chapter 7.

Consider the illustration of the function of one variable given in Figure 2.1.1. Points A and C correspond to local maxima, since by increasing or decreasing x starting from either of these two points the value of the function is decreased. At A the function has the global maximum in the interval $0 \le x \le 10$. Also points B and D are local minima and in this range D is the global minimum. The point E corresponds to a point of inflexion.

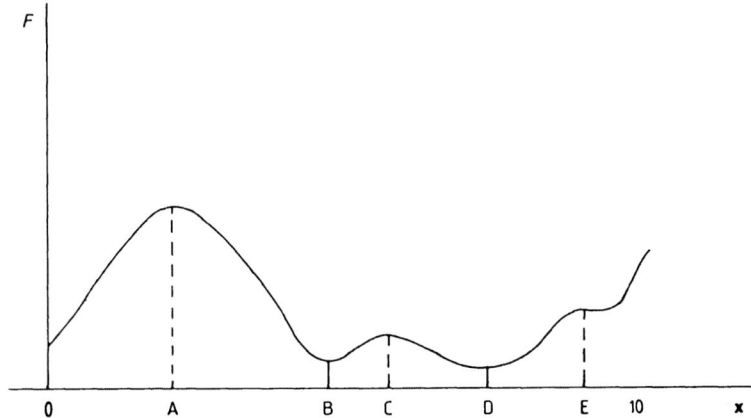

Figure 2.1.1 Local optima of a function of one variable.

The most powerful methods of optimisation require the function and its first and second derivatives to be continuous and we shall only deal with such functions.

There are very simple conditions that can be used to recognise local maxima and minima of such functions. These conditions are easily derived for functions of one variable.

Let us consider the function F in the neighbourhood of a local minimum x^*. For points x near x^* the Taylor series expansion of F is:

$$F(x) = F(x^*) + (x - x^*)(\frac{dF}{dx})_{x^*} + \frac{1}{2}(x - x^*)^2(\frac{d^2F}{dx^2})_{x^*}$$
$$+ \frac{1}{6}(x - x^*)^3(\frac{d^3F}{dx^3})_\xi \qquad (2.1.1)$$

where $x \leq \xi \leq x^*$.

In fact when x is sufficiently near x^* we can write:

$$F(x) - F(x^*) = (x - x^*)(\frac{dF}{dx})_{x^*} + \frac{1}{2}(x - x^*)^2(\frac{d^2F}{dx^2})_\varsigma \qquad (2.1.2)$$

where $x \leq \varsigma \leq x^*$.

Now for x sufficiently close to the local minimum the expression on the left hand side must be positive, or zero, as $F(x)$ cannot have a smaller value than $F(x^*)$. But the sign of the right hand side depends on whether x is larger or smaller than x^*. This inconsistency can only be resolved if

$$(\frac{dF(x)}{dx})_{x^*} = 0 \qquad (2.1.3)$$

which means that the slope of F is zero at a local optimum. Thus condition 2.1.3 is the necessary condition for x^* to be a maximum or a minimum. Points A, B, C, D and E in Figure 2.1.1 satisfy the condition. However, points B and D correspond to minima; point A and C correspond to maxima and E is called a point of inflexion. To discriminate between these points a condition involving higher derivatives is required.

Consider again equation 2.1.1. Because the first derivative is zero at x^*, the second derivative term becomes the dominant one (provided that it is not zero), so that

$$F(x) - F(x^*) = \frac{1}{2}(x - x^*)^2(\frac{d^2F}{dx^2})_{x^*} + \frac{1}{6}(x - x^*)^3(\frac{d^3F}{dx^3})_\xi \qquad (2.1.4)$$

To determine the condition for a minimum, we notice that since $(x - x^*)^2$ is always positive we will have

$$F(x) > F(x^*) \text{ for all } x \text{ sufficiently close to } x^*$$

if and only if

$$(\frac{d^2F}{dx^2})_{x^*} > 0 \qquad (2.1.5)$$

Similarly we can deduce that $F(x)$ has a maximum at x^* if condition 2.1.3 holds and

$$(\frac{d^2F}{dx^2})_{x^*} < 0 \qquad (2.1.6)$$

Should the second derivative be zero the point is a point of inflexion if the third derivative is non-zero.

In summary, the simplest conditions which characterise maxima and minima for functions of one variable are:

- The necessary condition for x^* to be a maximum or a minimum is

$$(\frac{dF}{dx})_{x^*} = 0$$

  Points satisfying this condition are called **stationary points**

- A condition for the stationary point x^* to be a minimum is

$$(\frac{d^2F}{dx^2})_{x^*} > 0$$

- A condition for the stationary point x^* to be a maximum is

$$(\frac{d^2F}{dx^2})_{x^*} < 0$$

- A condition for the stationary point x^* to be a point of inflexion is

$$(\frac{d^2F}{dx^2})_{x^*} = 0 \text{ and } (\frac{d^3F}{dx^3})_{x^*} \neq 0$$

When all first three derivatives are zero the same arguments can be extended to the next significant term in the expansion.

Example

Let us find and classify the stationary points of

$$F(x) = x^3(3x^2 - 5) + 1$$

The condition for the first derivative is

$$F'(x) = 15x^4 - 15x^2 = 15x^2(x - 1)(x + 1)$$

the stationary points are at

$$0, \ -1, \ 1$$

The second derivative is

$$F''(x) = 60x^3 - 30x$$

Thus

$$F''(-1) = -30 < 0, \text{ corresponds to a maximum}$$

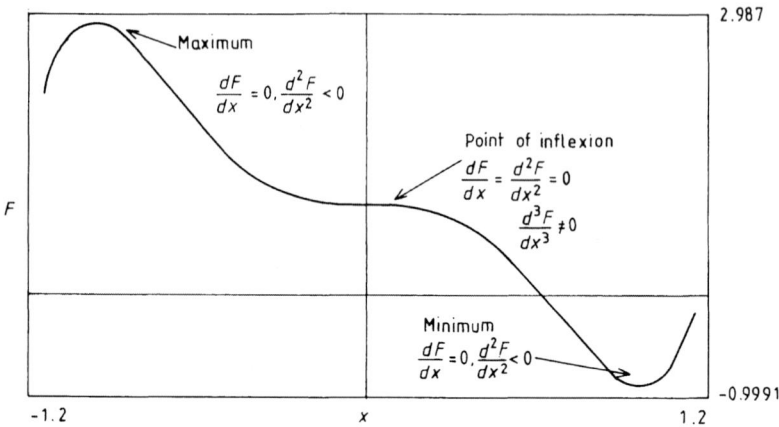

Figure 2.1.2 Stationary points of $F(x) = x^3(3x^2 - 5) + 1$.

$$F''(1) = 30 > 0, \text{ corresponds to a minimum}$$

$$F''(0) = 0, \text{ need to look at third derivative}$$

The third derivative is

$$F'''(x) = 180x^2 - 30$$

Then

$$F'''(0) = -30$$

Thus in the neighbourhood of $x = 0$, the function F increases for negative values of x and decreases for positive values of x. The point $x = 0$ corresponds to a point of inflexion. Figure 2.1.2 illustrates this.

When F depends on several variables $x_1, x_2, \ldots, x_n$, the same arguments can be applied to derive the appropriate conditions. In this case, for $\mathbf{x}$ sufficiently near $\mathbf{x}^*$, we have:

$$
\begin{aligned}
F(\mathbf{x}) - F(\mathbf{x}^*) \quad \approx \quad & (x_1 - x_1^*)(\frac{\partial F}{\partial x_1})_{x^*} \\
& + (x_2 - x_2^*)(\frac{\partial F}{\partial x_2})_{x^*} + \cdots \\
& + (x_n - x_n^*)(\frac{\partial F}{\partial x_n})_{x^*} \qquad (2.1.7)
\end{aligned}
$$

so, at a local minimum, for a non-positive left hand side of the above expression we must have:

$$\frac{\partial F}{\partial x_i} = 0 \; ; \quad i = 1, 2, \ldots, n \qquad (2.1.8)$$

The conditions that the second derivatives must satisfy for a minimum are rather more complicated than in the one-dimensional case, and we shall discuss them in Chapter 5.

However, here we state that all direct second derivatives

$$\frac{\partial^2 F}{\partial x_i^2}$$

must be positive, but the cross-derivatives

$$\frac{\partial^2 F}{\partial x_i \partial x_j}, \quad (i \neq j; i, j = 1, 2, \ldots, n)$$

need not be.

The zero first order derivatives at the optimum are a necessary condition. However there are other values of **x** which satisfy this condition but which are not local minima and the second derivative conditions are needed to discriminate between such points and local minima.

Example

Let us find the stationary points of

$$F(x_1, x_2) = 2x_1^2 + x_1 x_2^2 + x_2^2$$

The equations defined by the first partial derivatives are:

$$\frac{\partial F}{\partial x_1} = 4x_1 + x_2^2 = 0$$

$$\frac{\partial F}{\partial x_2} = 2x_1 x_2 + 2x_2 = 0$$

The solutions for this system of equations are:

$$(x_1 = 0, x_2 = 0), \; (x_1 = -1, x_2 = 2), \; (x_1 = -1, \; x_2 = -2)$$

The second order derivatives, $\partial^2 F/\partial x_1^2 = 4$ and $\partial^2 F/\partial x_2^2 = 2x_1 + 2$, are non-negative for these points, however the point $(0, \; 0)$ is the only

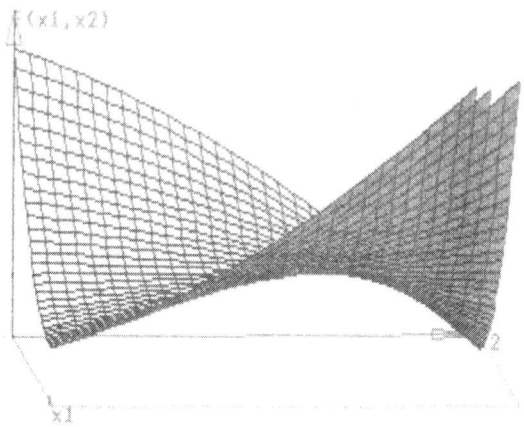

Figure 2.1.3 The shape of $F(x_1, x_2) = 2x_1^2 + x_1x_2^2 + x_2^2$ in the vicinity of the saddle point $(-1, 2)$.

one that corresponds to a minimum. The two other stationary points are called saddle points. For these points the function has a minimum along one direction and a maximum along the direction at right angles. Figure 2.1.3 illustrates the shape of F in the vicinity of the saddle point $(-1, 2)$.

It might appear that the problem of optimisation can be reduced to that of solving the system of nonlinear equations given by equations 2.1.8. This is indeed true, and when they can be solved using simple algebraic methods this may be the method to use. However in most cases the equations have to be solved numerically, and this turns out, in general, to be difficult. Very efficient methods of optimisation are not usually based directly on finding the solution of a system of equations. Most are in fact based on the simple idea of hill-climbing, in which the function F is analogous to the height of the hill, and $\mathbf{x}$ corresponds to the position of the climber who is on his way down and is looking for directions to reach $\mathbf{x}^*$, the bottom of the hill. To illustrate the idea of hill-climbing we shall consider the case of minimising functions of two variables.

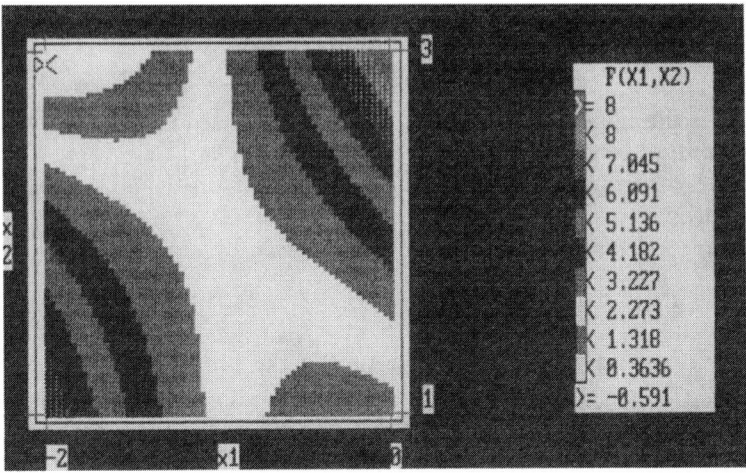

Figure 2.2.1 Contours of $F(x_1, x_2) = 2x_1^2 + x_1 x_2^2 + x_2^2$ in the vicinity of the saddle point $(-1, 2)$.

> ## 2.2 Contours of functions of two variables

We shall deal first with the graphical representation of functions of two variables. For these functions the vector $\mathbf{x}$ has only two components x_1 and x_2. One can think of the pair (x_1, x_2) as representing the coordinates of a point on this page. The value of F corresponding to the point can be thought of as the height above the page (depth below the page if F is negative). The set of points on the page which correspond to a fixed value of F define a set of lines on the page, each line being associated with a given height . These lines form a contour representation of the function F. Such maps of course are commonly used to represent height information over a two-dimensional surface such as a geographical area. Figure 2.2.1 shows the contour map of the function represented in Figure 2.1.3.

Example

Consider the simple function of two variables:

$$F(x_1, x_2) = (x_1 - 1)^2 + (x_2 - 1)^2$$

To obtain a contour of F equal to some selected value $k > 0$ we write

$$(x_1 - 1)^2 + (x_2 - 1)^2 = k$$

which is the equation of a circle of radius $\sqrt{k}$ centred at the point $(1, 1)$. Any contour of F is obtained by choosing the appropriate value for k.

The contours for this function thus happen to define a family of concentric circles.

A more interesting function to plot is given by

$$F(x_1, x_2) = (2x_1 + x_2) + (x_1^2 - x_2^2) + (x_1 - x_2^2)^2 \tag{2.2.1}$$

Now you should use the computer to plot the contours. Run program PREPARE, select option **E(dit** by typing E. Now type the function F as defined above using the conventions of BBC BASIC. For this case you need to type

```
(2*X1+X2) +(X1^2-X2^2) + (X1-X2^2)^2  <RETURN>
```

The range of values over which the plot is to be made needs to be specified. To tell the computer what your choice is, use the setting options.

For example, to plot the graph in the region

$$-2 \leq x_1 \leq 2 \qquad -2 \leq x_2 \leq 2$$

press 1 to select the option **x1(setting**, then press L to enter the lower limit and type your choice. The screen also displays a current value, -2 in this case, and the range of allowed values for L. If you just press **RETURN** the setting defaults to the current value. For the upper limit, enter H and proceed as before.

The **R(esolution** option determines the resolution of the plot and as for the other options the default value and the set of possible values are given on the screen when R is selected.

Pressing **RETURN** changes the list of options so that you can choose the range for x_2. Type 2 to set the range for x_2 and proceed as before. From the main menu select option **D(isplay**; a new list appears. To plot the graph select the option **S(urface**. The picture starts building on the screen.

It is possible to store a picture shown on the screen. To do it, go back to the main list of options by pressing **RETURN**.

You can scroll along this list with the arrow keys. To store the picture, select the option **S(ave**, and name the file on which the picture will be stored. The function given in equation 2.2.1 and its contours are already stored on your disc in the file named BEDP.

To retrieve a picture stored in a file, select **L(oad**, type the file name followed by **RETURN**. **Q(uit** stops the program.

We suggest you try several regions and from the picture of the contours visualise the shape of the surface. Try to draw an artist's impression of the surface. For example, if you choose the range $(-2, 2)$ and $(-2, 2)$ for the two independent variables, the contours will be kidney shaped. The surface can be visualised by imagining it as if "coming out" of the screen. The cross in the lower third of the contour plot, which represents the lowest computed value of the function, can be imagined at a distance greater than or equal to 1.09 units into the screen. The differently shaded regions represent points on the surface at distances from the screen equal to the number of units given in the table beside the contour plot. The resulting surface could be viewed as having the shape of a hospital-type bedpan and from now on we will refer to this function as BEDP.

A perspective of a three-dimensional plot for a surface can be obtained using the program 3D-PLOT. You can activate 3D-PLOT any time after building a function with PREPARE.

To see what your surface might look like in a three-dimensional space you run 3D-PLOT and then choose **D(isplay** followed by **S(urface**. A perspective of BEDP will start forming on the screen.

> 2.3   Directions on the contour map

The position of a point on the contour map is defined by a pair of numbers, say $\mathbf{P} = (2, 1)$. To plot $\mathbf{P}$ on the contour map of BEDP use the program CONTOUR; remember that PREPARE must have been used previously to load the function. Now activate the **D(isplay** option to show the contours of the function and select the option **P(oint**, you will be then prompted for the coordinates of the point you wish to plot. The pair of numbers which define a point could also be thought of as a vector stretching from the origin. The point $(2, 1)$ when considered as a vector would be expressed as

$$\mathbf{P} = \begin{bmatrix} 2 \\ 1 \end{bmatrix}$$

Try plotting some points for yourself.

Now suppose that the point **P** has been defined on the contour map. We want to see the effect of adding a vector **Q** to **P**.

Let

$$\mathbf{Q} = \left[\begin{array}{c} -0.5 \\ 0.5 \end{array} \right]$$

then:

$$\mathbf{P} + \mathbf{Q} = \left[\begin{array}{c} 1.5 \\ 1.5 \end{array} \right]$$

Let us see what the vector **P + Q** looks like on a contour plot. Enter the point **P**; now all vectors will stretch from **P**, so select the vector option and enter **Q**. (At this stage ignore the top window in the picture). We can interpret the new vector **P + Q** as a translation of the point defined by **P** to a new point specified by **P + Q**. The distance and direction in which the translation of **P** occurs is determined by the choice of **Q**.

Exercise

Plot vectors **P + Q** for the following vectors for **Q**:

$$\left[\begin{array}{c} 1 \\ 1 \end{array} \right], \left[\begin{array}{c} 1 \\ -1 \end{array} \right], \left[\begin{array}{c} -1 \\ 0 \end{array} \right]$$

Now let us see the effect of multiplying **Q** by a scalar s:

$$s\mathbf{Q} = \left[\begin{array}{c} sq_1 \\ sq_2 \end{array} \right]$$

Multiplying a vector by a scalar is the same as multiplying each component of the vector by the scalar.

Choose several values of s and, for some choice of **Q**, work out $s\mathbf{Q}$ using pencil and paper and display **P + $s\mathbf{Q}$** on the contour map. You will see that multiplying **Q** by s does not modify the direction of **Q**, but it does change the distance that the point moves. Thus $s\mathbf{Q}$ defines the set of points along the direction **Q**.

> *Any point on a straight line passing through a point* **P** *in the direction* **Q** *can be specified by a suitable choice of s in the expression* $\mathbf{x} = \mathbf{P} + s\mathbf{Q}$.

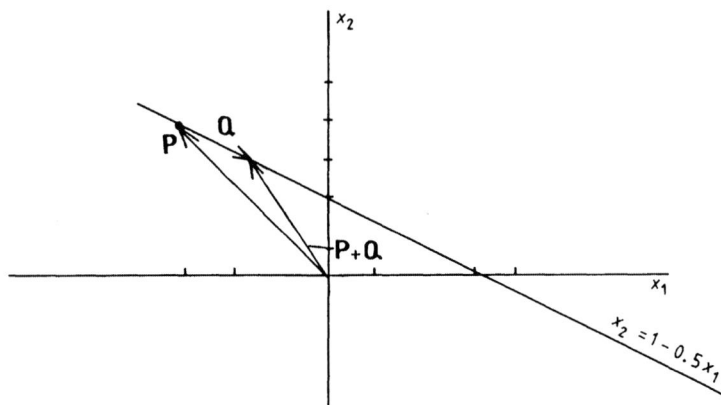

Figure 2.3.1 Vector representation of a straight line.

Example

The vector form of the line through the point

$$\mathbf{P} = \left[\begin{array}{c} -1.8 \\ 1.9 \end{array} \right]$$

in the direction of the vector

$$\mathbf{Q} = \left[\begin{array}{c} 1.0 \\ -0.5 \end{array} \right]$$

is

$$\mathbf{x} = \left[\begin{array}{c} x_1 \\ x_2 \end{array} \right] = \left[\begin{array}{c} -1.8 \\ 1.9 \end{array} \right] + s \left[\begin{array}{c} 1 \\ -0.5 \end{array} \right] \qquad (2.3.1)$$

Figure 2.3.1 shows the line defined by the vector **x**.

Let us find the equation of this straight line in the more familiar form $x_2 = mx_1 + n$.

Equation 2.3.1 can be expressed as two equations for the components of **x**, thus:

$$\begin{array}{rcl} x_1 & = & -1.8 + s \\ x_2 & = & 1.9 - 0.5s \end{array}$$

Eliminating s between these two equations, we obtain first:

$$s = x_1 + 1.8$$

And thus:

$$x_2 = 1 - 0.5x_1$$

> 2.4   Directions to the optimum

We now know how to specify a direction on the contour map. We will
use this knowledge to develop the methodology for optimisation. When
the function F is specified at some point $\mathbf{x} = \mathbf{P} + s\mathbf{Q}$, i.e. somewhere
along the line through a given point $\mathbf{P}$ in a given direction $\mathbf{Q}$, the only
independent variable in F is the scalar s. Thus $F(\mathbf{P} + s\mathbf{Q})$ is only a
function of s. It represents the shape of the cross-section of the contour
map when sliced along $\mathbf{P} + s\mathbf{Q}$. Let us use the computer to illustrate the
search for the minimum of the function F along the line:

$$\mathbf{x} = \mathbf{P}^{(0)} + s\mathbf{Q}^{(0)} = \begin{bmatrix} -1.8 \\ 1.9 \end{bmatrix} + s \begin{bmatrix} 1 \\ -0.5 \end{bmatrix} \qquad (2.4.1)$$

and we can think of $\mathbf{Q}^{(0)}$ as a 'search direction'.

We need first to use CONTOUR to display the contours of BEDP,
then plot the point $\mathbf{P}^{(0)}$ by typing P to select the option $\mathbf{P}$(oint and
enter the values of the coordinates x_1 and x_2 when prompted. To show
the search direction $\mathbf{Q}^{(0)}$ select the option $\mathbf{V}$(ector by typing V and in
response to the prompts type the components of $\mathbf{Q}^{(0)}$.

A line will appear on the map, with one end at $(-1.8, 1.9)$, marked
by a cross and the other end at $\mathbf{P}^{(0)} + \mathbf{Q}^{(0)}$. The points in between
correspond to the end of the vector $\mathbf{x}$ when $0 \le s \le 1$.

The graph of $F(s)$ against s is shown in the window above the contour
plot. We can extend the line defined by $\mathbf{x}$ by simply choosing s greater
than one or move the line in the opposite direction with values of s less
than zero.

However in CONTOUR we normalise s to take values between zero
and one so to extend the line we simply replace $\mathbf{Q}^{(0)}$ by a multiple of
itself. For example, the vector $4\mathbf{Q}^{(0)}$ with components $(4, -2)$ extends
the line to the edge of the map, as you can verify by plotting it. The
graph in the upper window shows $F(s)$ with s between 0 and 1 where
$F(0) = F(\mathbf{P}^{(0)})$ and $F(1) = F(\mathbf{P}^{(0)} + 4\mathbf{Q}^{(0)})$. The position of the
minimum is marked by a vertical bar at $s = 0.4844$ and the value of $F(s)$

at this point is shown above the window to be 0.8904. The minimum of F along this line thus occurs at the point:

$$\mathbf{X} = \mathbf{P}^{(0)} + 0.4844 \times 4\mathbf{Q}^{(0)}.$$

This may not be the position of the local minimum of F, but we can accept it as a better estimate than $\mathbf{P}^{(0)}$ of the position of the local minimum because $F(\mathbf{X}) < F(\mathbf{P}^{(0)})$.

In the hill-climbing analogy, the climber who starts his descent at $\mathbf{P}^{(0)}$ will be going down the hill when he moves along the direction $\mathbf{Q}^{(0)}$ until he reaches $\mathbf{X}$. Now he needs to change direction if he wants to progress in his descent.

So let us move the reference point to $\mathbf{X}$ (think of it as $\mathbf{P}^{(1)}$); the program keeps the coordinates of the optimum of $F(s)$ as the default value for the option $\mathbf{P}$(oint. To enter $\mathbf{X}$, press P followed by **RETURN** twice. The program prompts you to choose whether to retain the previous vector or not, select yes so you can see the path followed.

To improve on the new estimate $\mathbf{P}^{(1)}$ we can repeat the process.

A promising direction to search along next might be south-westerly on the screen. Let us try the direction

$$\mathbf{Q}^{(1)} = \left[\begin{array}{c} -0.1 \\ -1.0 \end{array} \right]$$

Enter $\mathbf{Q}^{(1)}$. The line displayed is too short. Try $3\mathbf{Q}^{(1)}$. This was a good guess! The search direction passes very near the optimum. We can now have $\mathbf{P}^{(2)} = \mathbf{P}^{(1)} + 0.5938 \times 3\mathbf{Q}^{(1)}$ and iterate again. So enter $\mathbf{P}^{(2)}$ using $\mathbf{P}$(oint and pressing **RETURN** twice. If we choose $\mathbf{Q}^{(2)}$ to be a vector pointing left and a bit upwards, for example

$$\mathbf{Q}^{(2)} = \left[\begin{array}{c} -1.0 \\ 0.1 \end{array} \right]$$

we obtain a line passing through the minimum, which occurs at the point

$$\mathbf{P}^{(3)} = \mathbf{P}^{(2)} + 0.031\mathbf{Q}^{(2)} = \left[\begin{array}{c} -0.1344 \\ -0.8406 \end{array} \right], \text{ with } F(\mathbf{P}^{(3)}) = -1.091.$$

Let us review the procedure to obtain $\mathbf{P}^{(3)}$. We started with an arbitrary point $\mathbf{P}^{(0)}$ and a direction $\mathbf{Q}^{(0)}$, which was chosen arbitrarily for the purpose of illustration. We stepped along $\mathbf{Q}^{(0)}$ a distance specified by the value of s which minimised F along that direction. This determined $\mathbf{P}^{(1)}$. Starting at $\mathbf{P}^{(1)}$ we continued the search along a new promising direction $\mathbf{Q}^{(1)}$ until we found $\mathbf{P}^{(2)}$, and so on until a point close to the required local minimum was obtained. The procedure thus consisted of generating a sequence of promising directions and moving along each in turn to a better point. This simple idea forms the basis of most iterative methods of optimisation.

Looking at contours and one-dimensional graphs has helped us to visualise the line search procedure for optimisation but is a most inefficient approach since the function needs to be computed at many points in order to generate a contour plot, and most of these points are far away from both the minimum and the trial points and so are wasted. In addition, of course, such a method is impossible if there are more than two independent variables.

In practice therefore there is no contour plot to inspect, and indeed if there were, the above procedure would be unnecessary as the position of the minimum is shown in the plot. Instead, there is only the knowledge of the value of the function and possibly of its derivative at each trial point $\mathbf{P}^{(0)}, \mathbf{P}^{(1)}$, etc. and we need to develop methods of using this information to generate good search directions and algorithms to find the minimum along a search direction without calculating the function F too many times. In fact the measure of efficiency of an optimisation method is usually the number of function evaluations which are required.

Exercise

Use the computer programs to reproduce the pictures shown in Figures 2.1.2 and 2.2.1.

> 2.5 Summary

In this chapter we have introduced the essential ideas involved in the construction of an optimisation algorithm. We can identify at this stage two aspects of this approach which are most demanding of computing resources. They are:

- finding the search direction
- determining the size of the step along the search direction.

We shall deal first, in the next chapter, with the problem of determining the step size along a given direction.

> Chapter 3

> Line Searches

> 3.1 Introduction

The previous chapter discussed the idea of minimising a function of several variables using the repeated minimisation of a function along search directions.

We showed that the problem of minimising a function of many variables $F(\mathbf{x})$ could be transformed into that of the repeated minimisation of a function $F(\mathbf{P}+s\mathbf{Q})$ of the single variable s. Therefore the procedure for finding the minimum of functions of one variable is an important element in any optimisation method. We deal in this chapter with such procedures.

> 3.2 Optimisation of functions of one variable

An approach to the problem of finding the minimum of the function of one variable $F(x)$ can be based on solving the equation

$$\frac{dF(x)}{dx} = 0 \qquad (3.2.1)$$

which is a necessary condition for a minimum. However for this method to be effective the derivative of $F(x)$ has to be available and the above equation must be simple to solve. We will now develop numerical methods which apply whether or not the above conditions hold.

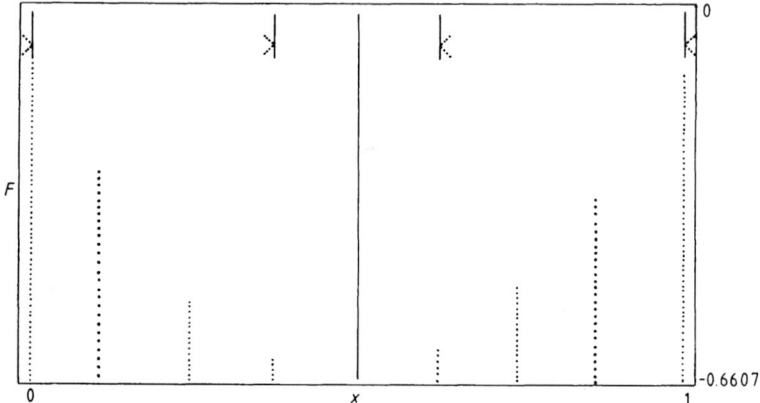

Figure 3.2.1 The grid method applied to $F(x) = x^2 - 3x \exp(-x)$.

> 3.2.1   The grid method

The simplest method of determining the minimum of F consists of finding an interval containing a single minimum and splitting it into a number of subintervals of equal size. The method is best described by an example.

Consider the function

$$F(x) = x^2 - 3x \exp(-x) \qquad (3.2.2)$$

which has a local minimum in the interval $0 \leq x \leq 1$ as shown in Figure 3.2.1. Let us divide this interval into, say, eight subintervals of equal size. For this we need nine function evaluations. Figure 3.2.1 shows $F(x)$ calculated at the nine points of interest. When $x = 0.5$, $F(x)$ takes the smallest value and thus the minimum of F lies in the interval $[0.375, 0.625]$.

We can repeat the process and divide this interval into eight equally spaced subintervals for which only six new function evaluations are required, as the values of F are known at the end points and at the centre. The procedure can be repeated until the solution is bracketed within the required accuracy.

This very simple method is generally inefficient because it produces too many wasted evaluations of the function and to specify the solution with some accuracy, using a manageable number of function evaluations, requires a fairly small interval to start with.

An improved grid method can be obtained by iteratively dividing the intervals into four equal parts. The first iteration uses five function evaluations and reduces the estimating interval by half; at each subsequent iteration the interval is reduced again by half but with only two extra function evaluations. To obtain the same accuracy as in the above example seven calculations of $F(x)$ are required against the nine used before. The option **S(earch E(qual** in the program SEARCH contains an implementation of the method. When selecting this option you will be prompted to set either the absolute or the relative error of the location of the optimum. The default value for these options is obtained by pressing the **RETURN** key.

To improve on the efficiency of the grid method we need to lift the restriction of dividing the original interval into equal size subintervals.

> ## 3.2.2   The Fibonacci search

Let us consider a function that has a single local minimum in the interval $[a, b]$, and for which the values of the function are known at a and b. How many additional evaluations of the function do we need in order to determine the position of the minimum with an error of size ϵ?

Let us try one evaluation of F at some point x_m contained in $[a, b]$. This gives extra information on F but is not enough for discriminating in which subinterval $[a, x_m]$ or $[x_m, b]$ the minimum might lie. The function might follow curve A or curve B, and with a single internal point x_m it is not possible to discriminate in which of the two possible intervals the minimum lies. Figure 3.2.2 illustrates the two possibilities for the position of the minimum.

However, an extra evaluation of F at a point x_n, located anywhere in $[a, b]$, would allow us to reduce the interval of uncertainty by selecting one of the two intervals $[a, x_n]$ or $[x_m, b]$. To select the interval which is certain to contain the minimum, we examine the function values and choose the one with a smaller function value at its internal point than at

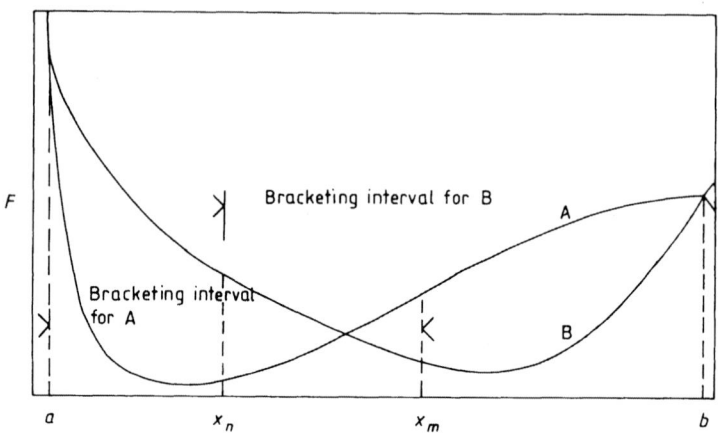

Figure 3.2.2 Bracketing the minimum.

its ends; we call it a bracketing interval. This is shown in Figure 3.2.2. Thus two internal points are the minimum number of points required to specify the bracketing interval in which the minimum is certain to lie.

Now we need to determine where in $[a, b]$ we should locate these internal points. Let us consider first where to locate x_n given that x_m was already chosen. Referring back to Figure 3.2.2, for a given selection of the position of x_n there are two possible bracketing intervals, one of length L_{N-1} when the minimum is in $[x_m, b]$ and the other of length R_{N-1} for the minimum in the interval $[a, x_n]$. Let us choose to locate x_n such that the length of the two possible intervals are the same. That is:

$$x_n - a = b - x_m$$

hence

$$x_m - a = b - x_n \qquad (3.2.3)$$

That is both points are at the same distance from the ends.

By locating x_n so that $R_{N-1} = L_{N-1}$ it is not possible to select one interval larger than the other.

We are therefore partitioning the bracketing interval so that we minimise the maximum possible error in locating the minimum. Any other partition with one of the intervals smaller than the other would give a

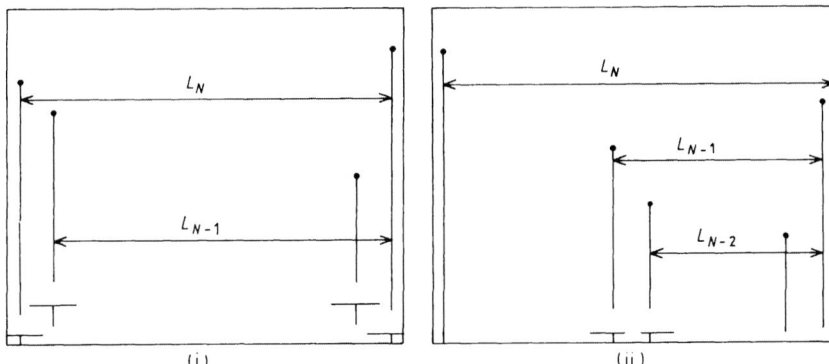

Figure 3.2.3 The internal point x_m in the bracketing interval. (i) near the end of the interval, (ii) near the middle of the interval.

risk of producing a larger maximum error. A criterion which leads to decision rules which minimise the maximum error is called **minimax**.

Let us suppose that the interval $[a, x_m]$ is found to contain the minimum. Then in the next iteration the points are relabelled:

$$a = a, \ b = x_m$$

and the new internal point is located according to equations 3.2.3. Clearly the sequence of internal points which are generated is fixed once x_m is chosen.

The question then arises: how do we choose x_m for an efficient minimax search? For example, if x_m is near the end of the interval then L_{N-1} is not much smaller than L_N so we gain little information. On the other hand for x_m near the middle of the interval we have a large reduction in the interval of uncertainty. However in this case x_n would be near x_m which leads to L_{N-2} having a value close to L_{N-1}. The total reduction of interval, over the two iterations, is not very great. Figure 3.2.3 illustrates this.

The problem of selecting x_m is one of obtaining the best reduction of uncertainty interval over the total number of iterations. Let us see how we can achieve this using an example. We wish to find the minimum of

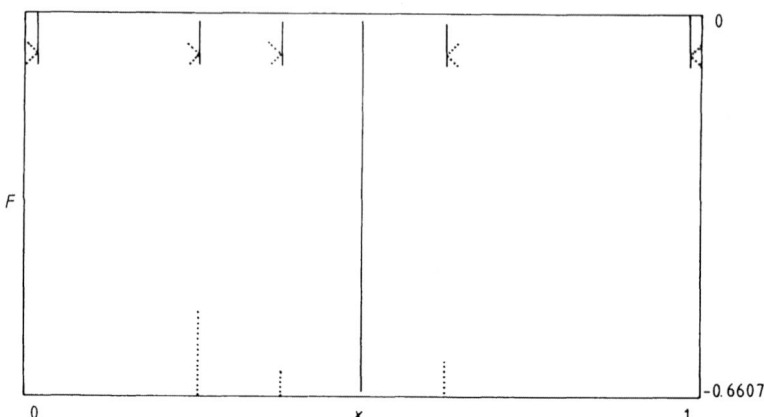

Figure 3.2.4 The Fibonacci search for $F(x) = x^2 - 3x \exp(x)$.

the function given in equation 3.2.2; let $L_N = [0,1]$. Let us set $x_m = 0.25$, then by symmetry $x_n = 0.75$. From the diagram in Figure 3.2.4 the interval containing the minimum is $L_{N-1} = [0, 0.075]$. By symmetry the next point is at 0.50 and the minimum now lies in the interval $L_{N-2} = [0.25, 0.75]$. The next internal point in the sequence is at 0.50. Here the iterations must end as the internal point in L_{N-2} is in the middle of the interval and the symmetry rule forces any new point to coincide with it. Still, L_{N-2} is subdivided into two equal intervals L_{N-3} but we cannot, using this method, discriminate between them for the position of the minimum. With this particular choice of x_m we have produced a sequence of intervals which seem to satisfy

$$L_N = L_{N-1} + L_{N-2} \qquad (3.2.4)$$

The expression in equation 3.2.4 is called a *recurrence relation of the second order* and defines a sequence of interval lengths. The first two numbers in the sequence, L_0 and L_1, need to be specified; once this has been done, all the elements in the sequence can be computed using equation 3.2.4.

The American mathematician Kiefer [1] proved, in the early fifties, the optimality of the minimax search when it is based on equation 3.2.4

and that the best choice for L_0 and L_1 is related to that of the sequence:

$$F_N = F_{N-1} + F_{N-2} \text{ , with } F_0 = F_1 = 1$$

which gives the numbers

$$1, 1, 2, 3, 5, 8, 13, 21, 34, 55, 89, 144, \ldots$$

This sequence is called the Fibonacci sequence after the medieval Italian mathematician.

To specify the searching algorithm we need a rule to determine the position of the first internal point in the interval $[a, b]$; the other points follow by symmetry. To find the position of the minimum with an error less than ϵ, we need to determine its estimate at the centre of the interval 2δ of size 2ϵ or less.

With an initial interval $L_N = [b, a]$ and a final one of size δ we can build the sequence of intervals:

$$\delta, \delta, 2\delta, 3\delta, 5\delta, \ldots, F_N\delta, \ldots$$

which can be seen to satisfy equation 3.2.4.

So by making

$$L_N = F_N\delta = b - a \qquad (3.2.5)$$

it follows that

$$\delta = \frac{(b-a)}{F_N}$$

Thus given a, b and ϵ the problem of determining the sequence of bracketing intervals is that of finding the smallest Fibonacci number that satisfies

$$\delta \leq \epsilon$$

The internal points are obtained by applying equation 3.2.5 to the interval L_{N-1}.

Thus

$$x_m = b - \delta F_{N-1}$$
$$x_n = a + \delta F_{N-1}$$

Example

Let us apply this to the problem of finding the minimum of $F(x) = x^2 - 3x \exp(-x)$ in the interval where

$$a = 0, \ b = \ 1 \text{ and } \epsilon = 0.15.$$

The smallest Fibonacci number for which

$$\delta = \frac{1}{F_N} \leq 0.15$$

is $F_5 = 8$, giving $\delta = 0.125$.

We therefore locate x_m at 5δ units from b and x_n at 5δ from a.

Using the option **S(earch F(ib** from the program SEARCH we obtained the iterations displayed in Table 3.2.1.

n	x_l	x_m	x_n	x_u	F_m	F_n
1	0.000	0.375	0.625	1.000	-0.633	-0.613
2	0.000	0.250	0.375	0.625	-0.522	-0.633
3	0.250	0.375	0.500	0.625	-0.633	-0.660

Table 3.2.1 Iterations for the Fibonacci search.

The minimum is found at $x = 0.500$ with an error of 0.125.

With the Fibonacci search we arrived at a final bracketing interval with an internal point at its centre, thus giving a maximum error of estimation equal to half the final bracketing interval. We have used $N - 1$ function evaluations ($5 - 1 = 4$) for a reduction of $2/F_N$ of the original interval.

Summary of the Fibonacci search algorithm

1. Determine the interval $[a, b]$ that contains the minimum.

2. Set the maximum error ϵ within which the minimum is to be located.

3. Find the smallest Fibonacci number F_N for which

$$\delta = \frac{(b - a)}{F_N} \leq \epsilon$$

4. Set

$$
\begin{aligned}
x_l &= a \\
x_m &= b - \delta F_{N-1} \\
x_n &= a + \delta F_{N-1} \\
x_u &= b \\
N &= N - 1
\end{aligned}
$$

5. For
$$F(x_m) > F(x_n)$$

If $N = 2$ the minimum is at x_m, the algorithm ends.
Else, the new set of points is

$$x_l, x_l + (x_n - x_m), x_m, x_n$$

For
$$F(x_m) < F(x_n)$$

If $N = 2$ the minimum is at x_n, the algorithm ends.
Else, the new set of points is

$$x_m, x_n, x_m + (x_u - x_n), x_u$$

6. Relabel the points
$$x_l, x_m, x_n, x_u$$

and repeat 5.

Exercise

- Using the Fibonacci search, how many function evaluations are required to determine the position of the maximum of

$$F(x) = \frac{x}{x^2 + 1}$$

in the interval $[0, 2]$, with an error in the location of the position of less than 0.001?

- Use the option **F(ib** to verify your result.

> ### 3.2.3 The golden section search

The Fibonacci search uses a fixed number of function evaluations to determine the position of the minimum with a given maximum error. There are situations where the maximum error is not useful for obtaining the required precision; for example, we may be required to obtain the position of the minimum correct to a given number of significant figures. In this case the number of iterations cannot be fixed initially so the iterations continue until the number of correct significant figures is obtained.

The golden section search is a simple efficient method that has the required property. It satisfies the minimax principle and is also based on the Fibonacci recurrence relation given by equation 3.2.4 for the length of the uncertainty intervals. It differs from Fibonacci in the way in which the location of the intervals is chosen. In the golden section the demand that the Fibonacci search has to be the optimum minimax method is replaced by the condition that at each iteration the intervals are reduced by the same factor, that is

$$\frac{L_{N-1}}{L_N} = \tau$$

Then from the recurrence relation in equation 3.2.3

$$\frac{1}{\tau} = 1 + \tau \ , \ \text{or} \ \tau^2 + \tau - 1 = 0$$

The value of interest is the positive root of the above equation:

$$\tau = \frac{(-1 + \sqrt{5})}{2} = 0.618033989$$

The intervals of uncertainty evolve as

$$
\begin{aligned}
L_{N-1} &= \tau L_N \\
L_{N-2} &= \tau L_{N-1} = \tau^2 L_N \\
\ldots &= \ldots \\
L_{N-k} &= \tau^k L_N
\end{aligned}
$$

Thus each iteration reduces the interval by $\tau = 0.618033989$ This is illustrated in Figure 3.2.5.

When the minimum is contained in the interval $[a, b]$ the internal points x_m, x_n, as displayed in Figure 3.2.3, are determined by setting the conditions

$$x_n - a = b - x_m + \tau(b - a)$$

thus

$$x_m = b - (b - a)\tau$$
$$x_n = a + (b - a)\tau$$

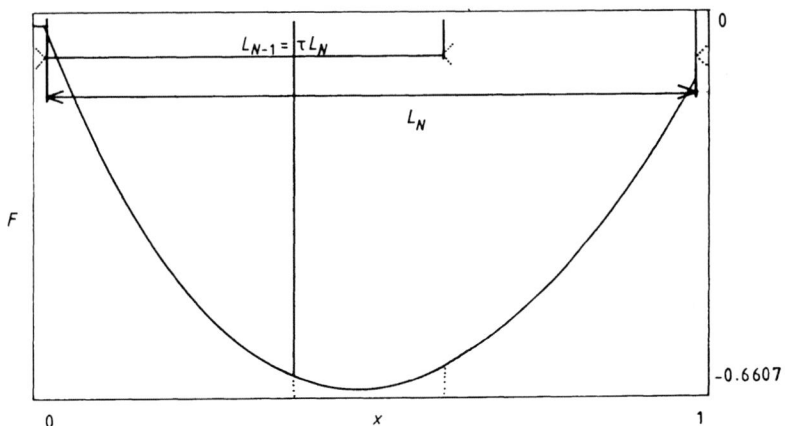

Figure 3.2.5 The golden section search.

Note that

$$L_N/L_{N-1} = L_{N-1}/(L_N - L_{N-1}) = \frac{1}{\tau} \qquad (3.2.6)$$

that is the ratio of the total to the largest interval is the same as the ratio of the largest interval to the smallest one. The ancient Greeks regarded this ratio as pleasing to the eye and called it the *golden section*. They used it in designing the proportions of their buildings.

The golden section can be slightly less efficient than the Fibonacci search, and it can be seen as a limiting form of that search. It can be shown that as N becomes large the ratio of the Fibonacci numbers satisfies

$$\lim_{N \to \infty} \frac{F_{N-1}}{F_N} = \tau \qquad (3.2.7)$$

For example, from the sequence given in equation 3.2.4

$$\frac{F_{10}}{F_{11}} = \frac{89}{144} = 0.61805555$$

not much different from τ. The largest reduction of interval in the Fibonacci search is $F_2/F_3 = 2/3 = 0.6666666$. Hence only for a few iterations is Fibonacci slightly better than the golden search. The simplicity of the golden search and its extra flexibility makes it a favourite.

The book by Harding and Quinney [2] discusses the topic of recurrence relations and their programs can be used to show the validity of the limit in equation 3.2.7

Summary of the golden section search algorithm

The function $F(x)$ has a minimum in the interval $[a, b]$.

1. Determine the interval $[a, b]$ that contains the minimum.

2. Let

$$
\begin{aligned}
x_l &= a \\
x_m &= b - (b - a)\tau \\
x_n &= a + (b - a)\tau \\
x_u &= b \\
F_m &= F(x_m) \\
F_n &= F(x_n)
\end{aligned}
$$

(note that F with a lower-case subscript denotes a function value, not a Fibonacci number!).

3. For

$$F_m < F_n$$

the new set of points is:

$$x_l, \ x_n - (x_n - x_0)\tau, \ x_m, \ x_n.$$

For

$$F_m > F_n$$

the new set of points is:

$$x_m, \ x_n, \ x_m + (x_u - x_m)\tau, \ x_u$$

4. Relabel the points as

$$x_l, \ x_m, \ x_n, \ x_u$$

5. Estimate relative error:

$$\text{If } \ F_m < F_n$$

Set
$$I = (x_u - x_m)/x_m$$
Else
$$I = (x_n - x_0)/x_n$$

If $I >$ required relative error, repeat step 3.

6.

If $F_m < F_n$

Optimum is x_m

Else

Optimum is x_n.

Exercises

- Show that the new internal points for the golden section search are as given in steps 1 and 2 of the algorithm.

- Prove equation 3.2.6.

Example

Let us use the golden section search to determine the position of the minimum of $F(x) = x^2 - 3x \, \exp(-x)$. The minimum is contained in the interval $[0, 1]$. Thus

$$x_m = 1 - (1 - 0)0.618034 = 0.381966$$

$$x_n = 0 + (1 - 0)0.618034 = 0.618034$$

Using the option S(earch G(old in the program SEARCH we obtained that after ten iterations the minimum lies in the interval $[0.4772, 0.4803]$. The minimum is located at $x = 0.48$, which is exact to two decimal places.

Exercises

1. Verify the results presented in the example above.

2. Consider the function of two variables discussed in Chapter 1:

$$F(x_1, \ x_2) = -(2x_1 + x_2) - (x_1^2 - x_2^2) - (x_1 - x_2^2)^2.$$

We wish to determine the minimum of F along the line

$$x = \left[\begin{array}{c} 1.8 \\ 1.8 \end{array} \right] + s \left[\begin{array}{c} 1 \\ -0.5 \end{array} \right]$$

  – How many iterations of the golden section search are required to find the minimum on the line with an error smaller than 0.001?
  – Use the option **G(old** to confirm your results.
  – Use the option **G(old** to find the minimum on the line exact to three significant figures.

> **3.2.4   The quadratic search**

The Fibonacci search was shown to be a way of locating the position of the minimum of a function with the smallest maximum error. Any other method, using the same number of function evaluations, has the chance of finding the minimum with a larger error. Nevertheless guarding against the worst happening, as in the Fibonacci search, usually means very slow progress.

One might wish to explore methods which in most occasions give a large reduction in the value of the function with very few function evaluations at the risk that some times they might be inefficient.

We will look at methods which usually determine very quickly good candidates for the position of the minimum.

The quadratic search algorithm is based on the idea that it might be effective to replace $F(x)$ by a quadratic function in a bracketing interval containing the position of the minimum The minimum of the quadratic is easy to calculate and is used as the estimated value for the minimum of $F(x)$. Using this new internal point a smaller bracketing interval can be determined and the process repeated until the change is sufficiently small.

Suppose that we know $F(x)$ at $x_l < x_n < x_u$ and that

$$F(x_l) > F(x_n) < F(x_u)$$

In this bracketing interval $F(x)$ is replaced by the quadratic polynomial

$$p(x) = a + bx + cx^2$$

As we know the values of $F(x)$ at three points, the values of a, b and c can be determined by solving the system of three linear equations with three unknowns:

$$p(x_l) = a + bx_l + cx_l^2 = F_l$$

$$p(x_n) = a + bx_n + cx_n^2 = F_n$$

$$p(x_u) = a + bx_u + cx_u^2 = F_u$$

The value x_m, of x, at which $p(x)$ attains its maximum is obtained by solving

$$\left(\frac{dp}{dx}\right)_{x_m} = b + cx_m = 0$$

hence

$$x_m = \frac{-b}{2c}$$

Solving the system of equations for b and c we find

$$x_m = \frac{1}{2}\frac{F_l(x_n^2 - x_u^2) + F_n(x_u^2 - x_l^2) + F_u(x_l^2 - x_n^2)}{F_l(x_n - x_u) + F_n(x_u - x_l) + F_u(x_l - x_n)}$$

This expression for x_m is unsuitable for numerical calculations when the bracketing interval is small, since then the points are close together and the differences of squares of very similar values involve a loss of significant figures. The following equivalent expression for x_m is better:

$$x_m = \frac{(x_l + x_u)}{2} \qquad (3.2.8)$$
$$- \frac{1}{2}\frac{(F_l - F_u)(x_u - x_n)(x_n - x_l)}{F_l(x_n - x_u) + F_n(x_u - x_l) + F_u(x_l - x_n)}$$

In this expression the major contribution to x_m is given by

$$\frac{(x_l + x_u)}{2}$$

which is the centre of the bracketing interval. The next term can be interpreted as a small correction. The new bracketing interval $[x_l, \ x_u]$ with the point x_n contained in the interval is specified by labelling the following set of points as x_l, x_n, x_u:

no change if

$$x_m = x_n$$

$x_n, \ x_m, \ x_u$ if $x_m > x_n$ and $F(x_m) < F(x_n)$

$x_l, \ x_n, \ x_m$ if $x_m > x_n$ and $F(x_m) \geq F(x_n)$

$x_l, \ x_m, \ x_n$ if $x_m < x_n$ and $F(x_m) < F(x_n)$

$x_m, \ x_n, \ x_u$ if $x_m < x_n$ and $F(x_m) \geq F(x_n)$

Summary of the quadratic search algorithm

1. Given the maximum size of the termination error ϵ, the bracketing interval

$$[x_l, x_u]$$

and the internal point x_n with

$$F_l > F_n < F_u$$

2. The new internal point x_m is at the position of the minimum of the interpolating quadratic polynomial of F at x_l, x_n and x_u and is given by

$$x_m = \frac{(x_l + x_u)}{2} - \frac{1}{2} \frac{(F_l - F_u)(x_u - x_n)(x_n - x_l)}{(x_n - x_u)F_l + (x_u - x_l)F_n + (x_l - x_n)F_u}$$

3. Select new bracketing interval $[x_l, x_u]$ containing the minimum of F.

4. If $|x_m - x_n| \leq \epsilon$ stop.

   Else, relabel the points and repeat from step 2.

Example

To illustrate the use of the quadratic search let us find the minimum of

$$F(x) = x^2 - 3x \exp(-x)$$

using the bracketing interval $[0, 1]$. The option **S(earch Q(uad** in the program SEARCH is used to calculate the first three iterations. Table 3.2.2 displays them.

You should now repeat the calculation **without** plotting the function.

After three iterations the minimum is found at $x = 0.4829$. However, when you repeat this calculation, it will be clear from the picture on the screen that there is much uncertainty in the location of the actual minimum, which could be anywhere in the bracketing interval. Thus we have found a position for the minimum but its true value might be at the other end of the interval! If you now superimpose the plot of F on the screen (press **RETURN** to leave **S**(earch and select the **P**(lot option), it becomes evident that we have found a very exact value for the minimum. Without the plot we cannot know how precise the value is, but of course when a plot can be used the method is redundant.

n	x_l	x_n	x_u	F_n	x_m	F_m
1	0.000	0.500	1.000	−0.6590	0.521	−0.6568
2	0.000	0.500	0.5213	−0.6590	0.4856	−0.6606
3	0.000	0.4856	0.5	−0.660602	0.4829	−0.660643

Table 3.2.2 Iterations for the quadratic search.

> 3.2.5   The cubic search

Having introduced the quadratic search it is natural to inquire whether we could improve the efficiency of the search by using a cubic polynomial instead of a quadratic.

To replace $F(x)$ by the cubic polynomial

$$p(x) = a + bx + cx^2 + dx^3 \qquad (3.2.9)$$

we need to determine the four parameters (a, b, c, d). Hence we need to have four suitable pieces of information in the interval containing the minimum. It has been found that when the function and its first derivative is known at the ends of a bracketing interval, giving the four required values, the cubic search is efficient for determining a point with a much reduced value of the function.

Let us assume that the function and its first derivative are known at the points x_l and x_u. Without loss of generality we can assume that $x_l = 0$, as we can always make x_l the origin of the coordinates system.

By setting respectively the values of $p(x)$ and its derivative to the values of F and its derivative, which we call here F', at the points $x_l = 0$ and x_u we get the equations:

$$a = F_l$$

$$b = F_l'$$

$$a + bx_u + cx_u^2 + dx_u^3 = F_u$$

$$b + 2cx_u + 3dx_u^2 = F_u'$$

Solving for c and d gives

$$c = \frac{1}{x_u} \left(\frac{3(F_u - F_l)}{x_u} - (2F_l' + F_u') \right), \quad \text{and}$$

$$d = \frac{1}{x_u^2} \left(F_l' + F_u' - 2\frac{F_u - F_l}{x_u} \right)$$

Let

$$S = \frac{F_u - F_l}{x_u} \tag{3.2.10}$$

Then

$$c = \frac{1}{x_u}(3S - F_u' - 2F_l'), \quad \text{and} \tag{3.2.11}$$

$$d = \frac{1}{x_u^2}(F_l' + F_u' - 2S). \tag{3.2.12}$$

We wish to use the position of the minimum of $p(x)$ to estimate the position of the minimum of $F(x)$.

The stationary points of $p(x)$ must satisfy

$$b + 2cx + 3dx^2 = 0$$

The roots of this equation are

$$x_{\pm}^{(m)} = \frac{-c \pm \sqrt{c^2 - 3bd}}{3d} \tag{3.2.13}$$

where $x_+^{(m)}$ corresponds to the solution with the positive sign for the square root, and $x_-^{(m)}$ the one with the negative sign.

It is simple to show that the solution $x_+^{(m)}$ corresponds to the minimum of the cubic.

However, we should be careful using this solution. When $c > 0$ we might have in the numerator of equation 3.2.13 the difference of two similar numbers which leads to a loss of numerical accuracy. It is better to use in this case the equivalent, but numerically stable, expression:

$$x^{(m)} = \frac{b}{-c - \sqrt{c^2 - 3bd}} \tag{3.2.14}$$

Thus the minimum of the cubic occurs at

$$x^{(m)} = \frac{-c + \sqrt{c^2 - 3F_l'd}}{F_l' + F_u' - 2S}x_u, \text{ when } c < 0 \tag{3.2.15}$$

or

$$x^{(m)} = \frac{F_l'}{-c - \sqrt{c^2 - 3F_l'd}}x_u, \text{ when } c > 0 \tag{3.2.16}$$

The algorithm selects the new bracketing interval by setting

$$x_l = x^{(m)} \text{ when } F'_{x^{(m)}} < 0$$

or

$$x_u = x^{(m)} \text{ when } F'_{x^{(m)}} > 0 \tag{3.2.17}$$

These conditions are illustrated in Figure 3.2.6.

Summary of the cubic search algorithm

1. Determine the interval $[a, b]$ that contains the optimum. Label the points

$$x_l = a$$

$$x_u = b$$

Set

$$x_{offset} = x_l$$

$$x_l = 0$$

2. Set ϵ. The algorithm stops when the size of the absolute value of the derivative of F is smaller than ϵ.

3. Find a bracketing interval

$$[x_l, \ x_u]$$

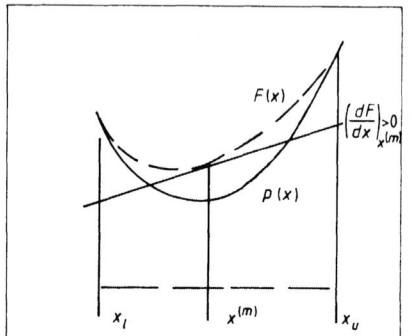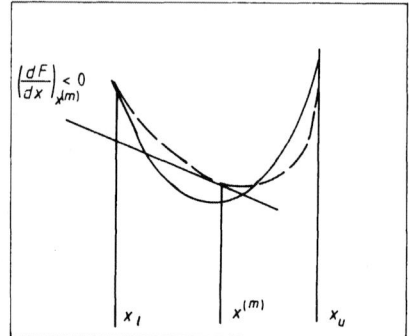

Figure 3.2.6 Bracketing the minimum in the cubic search.

4. Set

$$x_{offset} = x_{offset} + x_l$$
$$x_u = x_u - x_{offset}$$
$$x_l = 0$$

5. Find $x^{(m)}$ using equation 3.2.15 or equation 3.2.16.
6. Determine the new bracketing interval according to conditions 3.2.17.
7. If $|F'_{x^{(m)}}| < \epsilon$ stop.
   Else, repeat from 4.

Example

Let us use **S(earch C(ubic** to find the minimum of

$$F(x) = x^2 - 3x \exp(-x)$$

contained in the interval $[0, 1]$. When using this option you will be prompted for the largest acceptable value of the first derivative to terminate the algorithm, the default value is again obtained by pressing the return key.

The first iteration gives a very good solution; the derivative of F at this point is 0.00076. However, the next iteration gives a worse value for the position of the minimum, and after this there is slow convergence to the solution. This is typical of the method, which is normally used in cases where one application gives an acceptable solution. As with the quadratic search the method is inefficient for determining a small interval of uncertainty for the position of the minimum.

Exercise

Show that $x_+^{(m)}$ of equation 3.2.13 gives the position of the minimum of the cubic interpolating polynomial.

> 3.2.6 Finding the interval containing the minimum

The methods presented in this chapter assume that the minimum is located within a finite interval $[a, b]$. We discuss now a method to determine such an interval — that is, to bracket the minimum.

We calculate $F(x)$ at some point $a = x_l$ and start the search for the point b by evaluating $F(a + h)$, where h is a positive step size. If $F(a + h) > F(a)$ we are moving away from the minimum, it is therefore wise to turn around and calculate $F(a - h)$. If $F(a - h) > F(a)$, we have found a bracketing interval in $[a - h, a + h]$, if not we move further in the same direction to $a - 3h$, $a - 7h$, etc. doubling the step size each time we fail to establish an end for the bracketing interval. The method is illustrated in Figure 3.2.7.

The option **E(dit X(interval A(uto** can be used to determine a bracketing interval.

An alternative approach for bracketing the interval uses the derivative of the function and is generally applied with the cubic search. The method starts at a point x_l. Then if $F'_{x_l} < 0$ move a step h to $x_u = x_l + h$, otherwise relabel the point $x_u = x_l$ and move to $x_l = x_u - h$. If

$$F'_{x_u} > 0 \text{ or } F_{x_u} < F_{x_l}$$

the minimum lies in the interval $[x_l, x_l + h]$. This is illustrated in Figure 3.2.6.

The step h is varied as in the previous algorithm until the minimum is bracketed.

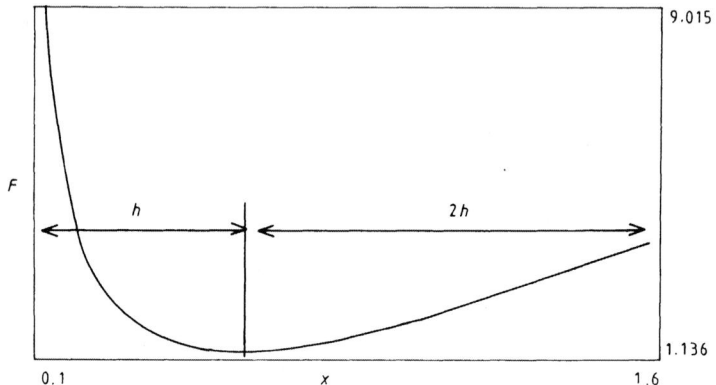

Figure 3.2.7 Bracketing the interval.

Exercise

Find the smallest positive value of x for which the function

$$F(x) = \frac{1}{x} - 3\cos x + 2$$

has a minimum, using:

- The grid method.
- The Fibonacci search.
- The golden section search.
- The quadratic search.
- The cubic search.

> 3.3 Summary

The methods described in this chapter can be divided into two groups.

The methods which are based on fixed interval reductions, such as Grid, Fibonacci and the Golden search, have a convergence rate which is independent of the type of function to be optimised so long as there is only one local minimum in the range considered. Fibonacci is the optimal minimax method, that is from all the minimax methods it is the one that will determine a specified bracketing interval for the minimum with the least number of function evaluations.

The quadratic and the cubic search seem to be efficient methods for most cases and are the ones generally used. However for very skewed functions, as for example that of the last exercise, these methods may be inefficient.

The quadratic and the cubic search are not based on an optimality principle (minimax) like the previous methods. Nevertheless they happen to be for most problems an efficient procedure for the determination of a point which has a much reduced value of the function.

We will see in Chapter 6 that in practice, for the optimisation of functions of several variables, exact optimisation along the line is not often used and all that is required to obtain the minimum of a function is to determine a point at which it has a sufficiently reduced value. It is for this purpose that the quadratic and cubic searches are most useful.

> Chapter 4

> Direct search methods

> 4.1 Introduction

The previous chapter dealt with methods to determine the position of the minimum along the search direction and now we turn to the problem of how automatically to generate good search directions when the only information available on the function are its values at a chosen starting point and at some subsequent trial points.

We will start by discussing the simplest possible method of generating search directions and will progress through different methods as the simpler ideas are modified to overcome their limitations.

> 4.2 Direct search methods

> 4.2.1 The univariate search algorithm

The simplest approach to generating search directions is the univariate search algorithm.

To begin with, no information is known about the function whose minimum we seek except its value at the starting point; there is therefore no reason to regard any direction of search as better than any other.

A methodical approach is to explore the effect of altering one variable at a time. For example in a function of two variables x_1 and x_2, one might keep x_2 constant and alter x_1 until an estimate of the minimum along this direction parallel to the x_1 axis is found. Starting from an initial point defined by the vector $\mathbf{x}^{(0)}$, a unit change in the variable x_1

is obtained by adding the vector $e^{(1)}$ to $x^{(0)}$, where $e^{(1)}$ is the two-dimensional vector

$$e^{(1)} = \begin{bmatrix} 1 \\ 0 \end{bmatrix} \qquad (4.2.1)$$

Thus the line passing through $x^{(0)}$ in the direction of increasing values of x_1 is given by

$$x^{(1)} = x^{(0)} + se^{(1)}$$

for any positive value of s. Negative values of s determine the portion of the line in the direction of decreasing values of x_1.

This easily generalises to the n-dimensional case, where

$$x^{(i)} = x^{(i-1)} + se^{(i)}, \quad \text{with} \ -\infty < s < \infty \qquad (4.2.2)$$

and

$$e^{(i)} = \begin{bmatrix} 0 \\ \vdots \\ 1 \\ \vdots \\ 0 \end{bmatrix} \qquad (4.2.3)$$

where the component equal to one is located in the i-th position.

The univariate search algorithm proceeds as follows.

Starting from $x^{(0)}$ the minimum of $F(x)$ is found along the direction $e^{(1)}$.

Let it be at the position given by

$$x^{(1)} = x^{(0)} + s_1 e^{(1)}$$

Now $x^{(1)}$ becomes the starting point for the next iteration. A one-dimensional search is again used to determine the distance s_2 along $e^{(2)}$ to define the position of the new point given by the vector

$$x^{(2)} = x^{(1)} + s_2 e^{(2)}.$$

When all the directions $e^{(i)}, i = 1, \ldots, n$, have been used then a cycle of the algorithm has been completed giving an estimate of the position of the minimum of $F(x)$ at the point defined by the vector $x^{(n)}$. In general this point will not be the actual minimum, so the cycle is repeated by setting $x^{(0)} = x^{(n)}$.

Let us experiment with this algorithm to determine the position of the minimum of the function BEDP of Chapter 1.

The function we wish to minimise is

$$F(\mathbf{x}) = (2x_1 + x_2) + (x_1^2 - x_2^2) + (x_1 - x_2^2)^2 \qquad (4.2.4)$$

Let us use the origin as the starting point, hence

$$\mathbf{x}^{(0,1)} = \begin{bmatrix} 0 \\ 0 \end{bmatrix}, \quad \text{which gives } F(\mathbf{x}^{(0,1)}) = 0.$$

Here the system for labelling the iterates has been changed to one in which the first superscript indicates the iteration number and the second the cycle number.

We search along a line parallel to the ordinate axis, then

$$\mathbf{x} = \begin{bmatrix} 0 \\ 0 \end{bmatrix} + s \begin{bmatrix} 1 \\ 0 \end{bmatrix} = \begin{bmatrix} s \\ 0 \end{bmatrix}$$

Substituting for $\mathbf{x}$ in $F(\mathbf{x})$ we obtain

$$F(s) = 2s(1 + s)$$

The minimum of $F(s)$ is at $s^* = -0.5$, with $F(s^{(*)}) = -0.5$.

To perform the iterations of the univariate search we will use the option U(nivar from the program DIRECT. So as usual load the function with PREPARE and then select DIRECT from the main menu. There are several options in DIRECT from which you can select D(isplay to show the contours of the function; do it. Selecting P(arameters allows you to change the stopping criteria, otherwise you can leave the program to operate with the default values. Selecting S(earch gives a new menu, the options in the new menu correspond to the algorithms while the option P(oint is for choosing where to start the iterations. The cross in the contour diagram shows the position of the present choice for P. One can find the coordinates of the present selection by activating P(oint. This will show a request for a coordinate and in the line below it a display of the present value for the coordinate and of the range of its allowed values. Pressing return keeps the existing value. In this case the present point is at $(0,0)$ so let us start the iterations with it.

The first search confirms the result above to give:

$$\mathbf{x}^{(1,1)t} = \begin{bmatrix} -0.5 & 0 \end{bmatrix}$$

$$F(\mathbf{x}^{(1,1)}) = -0.5$$

These quantities are represented in the screen by **xk** and **F new** respectively. The menu in this section of the program shows a **Z(oom** option index **Z(oom** option index. Press **Z** to locate the zooming area on the present selected point, press it again to activate the zoom.

For the second search therefore we start at the point $(0, -0.5)$ and the first cycle is completed by searching along the line

$$\mathbf{x} = \begin{bmatrix} -0.5 \\ 0 \end{bmatrix} + s \begin{bmatrix} 0 \\ 1 \end{bmatrix} = \begin{bmatrix} -0.5 \\ s \end{bmatrix}$$

The position of the minimum along the line is at

$$s^* = -0.63$$

then

$$\mathbf{x}^{(2,1)} \equiv \mathbf{x}^{(0,2)} = \begin{bmatrix} -0.5 \\ -0.63 \end{bmatrix}$$

and

$$F(\mathbf{x}^{(2,1)}) = 0.9725$$

At this stage one cycle of the algorithm is completed. We set out two cycles of this algorithm in Table 4.2.1 and display the steps in the univariate search in Figure 4.2.1. You can confirm the values in the table by proceeding with the iterations. Note that the window on the screen also shows with c the cycle number, with dx the modulus of the difference between the last two cycle points and df the difference between the values of the function at the last two cycle points.

The question now arises:

Should we do another cycle of the algorithm or is $\mathbf{x}^{(2,2)}$ an acceptable estimate of the position of the minimum?

The question could be readily answered if the first partial derivatives of $F(\mathbf{x})$ were known, because the partial derivatives evaluated at $\mathbf{x}^{(2,1)}$ must all be close to zero in order to satisfy a necessary condition for a minimum.

However the usefulness of direct search methods stems from the fact that they do not require the computation of derivatives. Derivatives are usually computationally expensive quantities to evaluate. In the absence of knowledge of the derivatives therefore we need to develop practical stopping rules.

(k,c)	$\mathbf{x}^{(k,c)}$	$\mathbf{e}^{(k,c)}$	s	$\mathbf{x}^{(k+1,c)} =$ $\mathbf{x}^{(k,c)} + s\mathbf{e}^{(k,c)}$	$F(\mathbf{x}^{(k+1,c)})$
0,1	$\begin{bmatrix} 0 \\ 0 \end{bmatrix}$	$\begin{bmatrix} 1 \\ 0 \end{bmatrix}$	-0.5	$\begin{bmatrix} -0.5 \\ 0 \end{bmatrix}$	-0.5
1,1	$\begin{bmatrix} -0.5 \\ 0 \end{bmatrix}$	$\begin{bmatrix} 0 \\ 1 \end{bmatrix}$	-0.63	$\begin{bmatrix} -0.5 \\ -0.63 \end{bmatrix}$	-0.9725
0,2	$\begin{bmatrix} -0.5 \\ -0.63 \end{bmatrix}$	$\begin{bmatrix} 1 \\ 0 \end{bmatrix}$	0.20	$\begin{bmatrix} -0.30 \\ -0.63 \end{bmatrix}$	-1.051
1,2	$\begin{bmatrix} -0.30 \\ -0.63 \end{bmatrix}$	$\begin{bmatrix} 0 \\ 1 \end{bmatrix}$	-0.10	$\begin{bmatrix} -0.30 \\ -0.73 \end{bmatrix}$	-1.079

Table 4.2.1 Iterations for the univariate search.

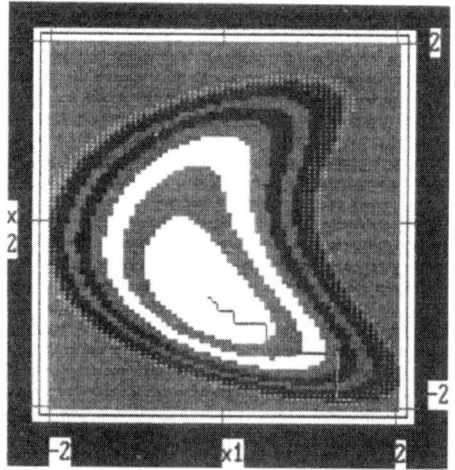

Figure 4.2.1 The univariate search.

For functions of n variables, we could have the following possible rules:

- Stop after the c-th cycle if

$$||\mathbf{x}^{(n,c)} - \mathbf{x}^{(0,c)}|| < \epsilon_1$$

where ϵ_1 is some small positive number chosen arbitrarily.

That is, stop if only a small improvement is made in the estimate of the position of the minimum during a cycle.

For the results given in Table 4.2.1 we have

$$||\mathbf{x}^{(2,2)} - \mathbf{x}^{(0,2)}|| = \left[(-0.30 + 0.5)^2 + (-0.73 + 0.63)^2\right]^{1/2}$$

which gives

$$||\mathbf{x}^{(2,2)} - \mathbf{x}^{(0,2)}|| = 0.224$$

Unless ϵ_1 is set to ≥ 0.224, a large value relative to the elements of $\mathbf{x}$, the cycles should continue.

- Stop after the c-th cycle if

$$|F(\mathbf{x}^{(n,c)}) - F(\mathbf{x}^{(0,c)})| < \epsilon_2$$

i.e. if F is decreased by only a small amount.

For the results in Table 4.2.1

$$|F(\mathbf{x}^{(2,2)}) - F(\mathbf{x}^{(0,2)})| = 1.079 - 0.9725 = 0.1065$$

Thus, unless an improvement of about 10% is considered acceptable the iterations should continue.

These criteria are easy to apply. However in choosing the ϵ's we should guard against the possibility of premature termination of the algorithm giving too imprecise a position of the minimum. Figures 4.2.1 and 4.2.2 illustrate some of the situations that frequently occur in optimisation.

The contours in Figure 4.2.1 show a case in which the univariate search is an effective algorithm, and either of the stopping criteria is appropriate.

Figure 4.2.2 however shows an example in which the algorithm converges very slowly to the minimum. In this case the univariate search was applied to the function

$$F(\mathbf{x}) = 100(x_2 - x_1^2)^2 + (1 - x_1)^2 \qquad (4.2.5)$$

which is called the Rosenbrock function. To plot the contours of the function we have used the option **H(igh F** in PREPARE with a value of 12.5 for the highest contour to be displayed.

It can be observed from the picture that this search is most ineffi-cient; starting at the point (0.4, 0.17) it has taken 28 cycles to reach the position at the upper right corner of the square drawn on the contours which corresponds to the point (0.57, 0.32). The steps of the search contained in the square and starting from its centre are shown in the picture at the right where the square has been expanded in the ratio of 4:1. The search is inefficient because for each cycle very small steps along each search direction are required to find the minimum along it.

Provided that one wishes to persist with the use of this algorithm, which in this case is so inefficient, then small values for the ϵ's should be chosen.

Exercise

The following exercises can be used to explore the behaviour of the univariate search algorithm when starting from different points and when using different values for ϵ_1 and ϵ_2.

1. Use **U(nivar** to find the number of iterations required to obtain the minimum of BEDP when $\epsilon_1 = \epsilon_2 = 0.01$.

2. Repeat the above exercise using the starting point:

$$\mathbf{x}^{(0,1)} = \begin{bmatrix} 1.0 \\ -1.5 \end{bmatrix}$$

> **4.2.2   The Davies, Swann and Campey (DSC) algorithm**

The last example illustrated the limitations of the univariate search al-gorithm. This method converges very slowly unless the contours are approximately of circular shape or line up with the coordinate direc-tions. The inefficiency of the algorithm can be put down to the fact that the search directions are always parallel to the coordinate axis and when the contours are elongated the moves provided by the univariate searches are not, in general, along an effective direction. The algorithm makes no use of the accumulating information about the function in order to improve its search strategy.

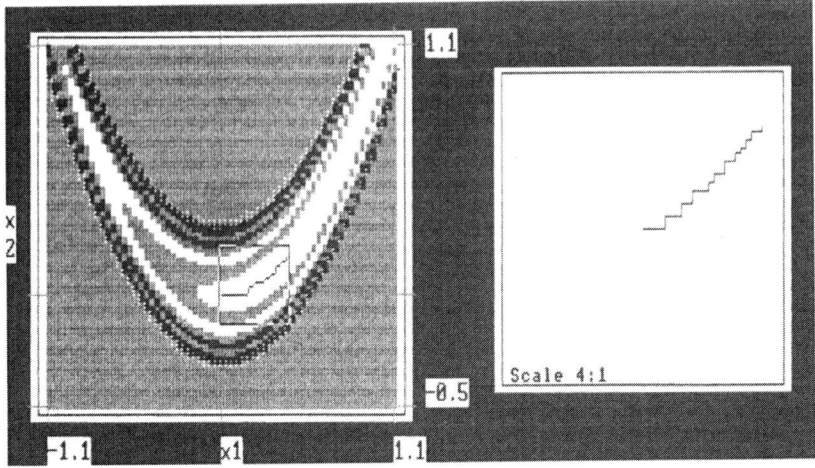

Figure 4.2.2 Inefficiency in the univariate search.

The DSC algorithm can be considered as a development of the univariate search method. It has a simple mechanism by which it adapts its search directions to take advantage of the accumulating information about the function. The algorithm can be illustrated using an example. The DSC algorithm starts from $x^{(0,1)}$ with one cycle of the univariate search to determine the point $x^{(2,1)} \equiv x^{(0,2)}$, where as before the first superscript indicates the iteration and the second one the cycle. In the next cycle, it searches along the direction defined by the line joining $x^{(0,1)}$ and $x^{(0,2)}$ to determine the point $x^{(1,2)}$. From $x^{(1,2)}$ it moves at right angles to the previous direction to end at the point $x^{(2,2)}$ and complete the second cycle of the algorithm. Now we can generate a new search direction by joining $x^{(0,2)}$ to $x^{(2,2)}$.

Figure 4.2.3 displays the cycles of the method when applied to the Rosenbrock function with the same starting point as that used for the univariate search. The method converges to the minimum at $(1, 1)$, shown by the cross, in 14 cycles. Here again the region contained in the square drawn over the contours has been expanded to show the steps of the algorithm and is displayed in the picture at the right.

We have described a simplified version of the DSC algorithm. In practice the algorithm is modified so that search directions are preserved unless a substantial change in the estimated position of the minimum is made during a cycle. In other words, if at the end of the c-th cycle

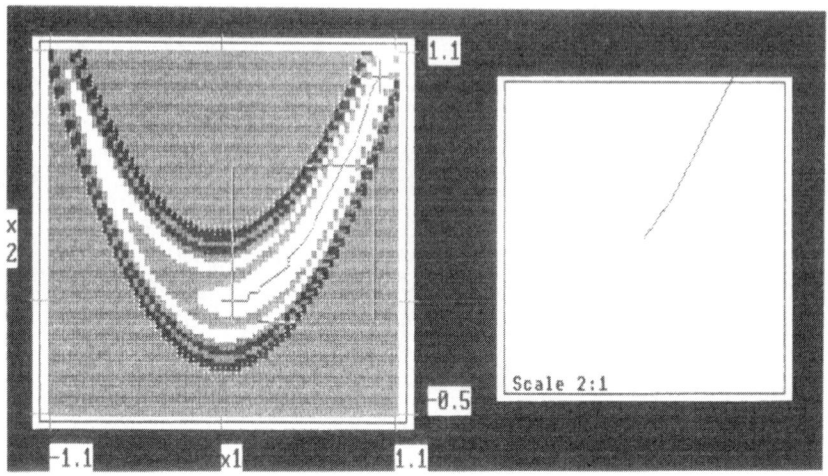

Figure 4.2.3 Iterations of the DSC algorithm.

the improvement in the determination of the position of the minimum
during that cycle is smaller than a certain specified quantity, then the
iterations continue from the last point but using the same directions as
in the previous cycle.

Let us apply this algorithm to the function defined by BEDP. We
can use as the first cycle the one obtained when we solved the problem
using the univariate search. Then

$$\mathbf{x}^{(0,1)} = \left[\begin{array}{c} 0 \\ 0 \end{array} \right]$$

and

$$\mathbf{x}^{(2,1)} \equiv \mathbf{x}^{(0,2)} = \left[\begin{array}{c} -0.5 \\ -0.63 \end{array} \right]$$

So the next search direction is

$$\mathbf{x}^{(0,2)} - \mathbf{x}^{(0,1)} = \left[\begin{array}{c} -0.5 \\ -0.63 \end{array} \right]$$

Thus we substitute

$$\mathbf{x} = \mathbf{x}^{(0,2)} + s \left[\mathbf{x}^{(0,2)} - \mathbf{x}^{(0,1)} \right] \text{ into } F(\mathbf{x})$$

and as before, find the value of s which maximises $F(s)$.

You might wish to do this as an exercise; we found $s = -0.095$. Hence

$$\mathbf{x}^{(1,2)} = \begin{bmatrix} -0.45 \\ -0.57 \end{bmatrix}, \; F(\mathbf{x}^{(1,2)}) = 0.9908$$

We now need to move along some direction

$$\mathbf{q} = \begin{bmatrix} q_1 \\ q_2 \end{bmatrix}$$

which is at right angles to the direction defined by

$$\begin{bmatrix} -0.5 \\ -0.63 \end{bmatrix}.$$

For these vectors to be orthogonal their scalar product must equal zero. Thus

$$-0.5q_1 - 0.63q_2 = 0.$$

Since the length of $\mathbf{q}$ is immaterial we can set q_2 arbitrarily to one. So

$$\mathbf{q} = \begin{bmatrix} -1.26 \\ 1.00 \end{bmatrix}$$

defines the required direction. (Note that the procedure assumes $q_2 \neq 0$; if it is, then q_1 will not be computable and we would instead set $q_1 = 1$.) In n-dimensions, a technique such as Gram–Schmidt orthogonalisation, which is fully described in reference [3], is used to generate the n orthogonal components of $\mathbf{q}$.

The cycle is completed by substituting

$$\mathbf{x} = \mathbf{x}^{(1,2)} + s\mathbf{q}$$

in $F(\mathbf{x})$. The value of s which maximises $F(s)$ is found to be $s = -0.225$. Then

$$\mathbf{x}^{(2,2)} = \begin{bmatrix} -0.17 \\ -0.80 \end{bmatrix}, \; F(\mathbf{x}^{(2,2)}) = 1.0948$$

To measure the rate of progress of the algorithm we compute the usual quantities defined. That is

$$||\mathbf{x}^{(2,2)} - \mathbf{x}^{(0,2)}|| = \left[(-0.17 + 0.45)^2 + (-0.80 + 0.57)^2 \right]^{1/2}$$

(k, c)	$\mathbf{x}^{(k,c)}$	$\mathbf{q}^{(k)}$	s	$\mathbf{x}^{(k+1,c)} = $ $\mathbf{x}^{(k,c)} + s\mathbf{q}^{(k)}$	$F(\mathbf{x}^{(k+1,c)})$
0,1	$\begin{bmatrix} 0 \\ 0 \end{bmatrix}$	$\begin{bmatrix} 1 \\ 0 \end{bmatrix}$	-0.5	$\begin{bmatrix} -0.5 \\ 0 \end{bmatrix}$	-0.5
1,1	$\begin{bmatrix} -0.5 \\ 0 \end{bmatrix}$	$\begin{bmatrix} 0 \\ 1 \end{bmatrix}$	-0.63	$\begin{bmatrix} -0.5 \\ -0.63 \end{bmatrix}$	-0.97
0,2	$\begin{bmatrix} -0.5 \\ -0.63 \end{bmatrix}$	$\begin{bmatrix} -0.5 \\ -0.63 \end{bmatrix}$	-0.09	$\begin{bmatrix} -0.45 \\ -0.57 \end{bmatrix}$	-0.9908
1,2	$\begin{bmatrix} -0.45 \\ -0.57 \end{bmatrix}$	$\begin{bmatrix} -1.26 \\ 1.00 \end{bmatrix}$	-0.225	$\begin{bmatrix} -0.17 \\ -0.80 \end{bmatrix}$	-1.0948

Table 4.2.2 Iterations of the DSC algorithm.

Then

$$\|\mathbf{x}^{(2,2)} - \mathbf{x}^{(0,2)}\| = 0.36$$

and

$$|F(\mathbf{x}^{(2,2)}) - F(\mathbf{x}^{(0,2)})| = 1.0948 - 0.9908 = 0.1040$$

These are large changes in $\mathbf{x}$ and F relative to the values of the quantities themselves and it would be reasonable to continue with the iterations. The iterations for the first cycle of the algorithm are set out in Table 4.2.2.

Exercise

- Perform another iteration of DSC for the above example.

- Determine the position of the minimum of BEDP using the method of DSC with starting point

$$\mathbf{x}^{(0)} = \begin{bmatrix} 1 \\ -1.5 \end{bmatrix}$$

- Use the option **D(SC** to find the minimum of the function which is shown in Figure 4.2.2:

$$F(\mathbf{x}) = 100(x_2 - x_1^2)^2 + (1 - x_1)^2$$

in the interval

$$-1.5 < x_1 < 1.5, \ -0.5 < x_2 < 1.5$$

with initial point at $(-1, 1.2)$.

For this exercise it is interesting to plot the contours of the function before doing the optimisation. When plotting the contours select the option **H(igh** and set it to a value of 20 in order to show up the deep valleys of the function.

It only remains to state the algorithm for functions of n variables as follows:

1. Set $c = 1$.
   Select some arbitrary starting point $\mathbf{x}^{(0,c)}$.

2. Carry out one cycle of the univariate search algorithm to produce $\mathbf{x}^{(n,c)} \equiv \mathbf{x}^{(0,c+1)}$.

3. Select $\mathbf{q} = \mathbf{x}^{(n,c)} - \mathbf{x}^{(o,c)}$ as a new search direction.

4. Generate $n-1$ directions mutually orthogonal and orthogonal to $\mathbf{q}$.

5. Compute a new set of points

$$\mathbf{x}^{(i,c+1)}, \quad i = 1, \ldots, n$$

   by searching, in turn, along $\mathbf{q}$ and each of the other $n - 1$ mutually orthogonal directions. Each search begins at the end point of the previous one.

6. If stopping criteria satisfied stop, else set $c = c + 1$ and repeat 3.

There are other methods which share with DSC the idea of determining search directions on the basis of information gained at successive points during the iteration. The method of Hooke and Jeeves [4] uses a cycle with two components, an exploratory phase and a pattern move.

In the exploratory phase the algorithm starts at a point $\mathbf{x}^{(i)}$ and it explores the possibility of a better point for the function by moving a fixed step h along directions parallel to the coordinate axis.

When a better point $\mathbf{x}^{(i+1)}$ is found in this phase a pattern move of the same fixed distance h is made along the direction $\mathbf{x}^{(i+1)} - \mathbf{x}^{(i)}$ to the new point $\mathbf{x}^{(i+2)}$ and the cycle is repeated.

The method uses for each move a fixed step rather than a line search.

Rosenbrock's method [3] shares with Hooke and Jeeves, the moves by fixed steps. It chooses to rotate the axis when a search along every direction parallel to the present coordinate system fails to find a better point for the function.

> 4.2.3   The simplex method

This chapter introduced two direct search methods. The univariate search, which is an obvious approach to solving the problem, and the more efficient method of Davies, Swann and Campey.

There is a method which uses a completely different strategy to determine the sequence of one-dimensional search directions.

The Simplex method[1] is associated with the names of Nelder and Mead and also those of Spendley, Hext and Himsworth. We will discuss it in general terms with illustrations for a function of two variables.

It proceeds by setting up a 'simplex' of points, i.e. a set of $n + 1$ points enclosing a volume in n-dimensional space . Thus when $n = 2$, the three points selected must not lie on a straight line; when $n = 3$, the four points chosen must not lie on a plane. The function is evaluated at every point in the simplex.

The method generates a new point at which the function is evaluated; the point with the largest value of the function is rejected and the new point is introduced into the simplex. In this way a sequence of simplices is created and the best point in the sequence converges to a local minimum.

The simplex algorithm is as follows:

Consider $n + 1$ representing the first simplex.

1. The points are labelled such that

$$F(\mathbf{x}^{(l)}) \leq \cdots \leq F(\mathbf{x}^{(s)}) \leq F(\mathbf{x}^{(h)})$$

---

[1] Do not confuse with the simplex method in linear programming!

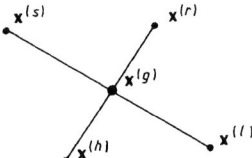

Figure 4.2.4 The reflection in the simplex method.

2. The convergence criterion for the algorithm is that the new simplex satisfies:

$$\sum_{i=1}^{n+1} \left(F(\mathbf{x}^{(i)}) - F(\mathbf{x}^{(l)}) \right) < \epsilon \qquad (4.2.6)$$

This criterion establishes a simplex having a deviation from the best obtained value less than a specified positive value ϵ which should be selected taking account of the value of F.

If equation 4.2.6 is satisfied accept $\mathbf{x}^{(l)}$ as the minimum and stop.

3. A new point is generated by reflecting the worst point $\mathbf{x}^{(h)}$ about the point

$$\mathbf{x}^{(g)} = \frac{1}{n} \sum_{i \neq h} \mathbf{x}^{(i)}$$

which is the centroid of all the points excluding the worst. Figure 4.2.4 illustrates the reflection operation for $n = 2$. The reflected point $\mathbf{x}^{(r)}$ satisfies

$$\mathbf{x}^{(h)} - \mathbf{x}^{(g)} = \mathbf{x}^{(g)} - \mathbf{x}^{(r)}$$

A better point than $\mathbf{x}^{(s)}$ is sought along the line $\mathbf{x}^{(h)} - \mathbf{x}^{(r)}$. So when

$$F(\mathbf{x}^{(r)}) > F(\mathbf{x}^{(s)})$$

means that we may have moved out of the simplex with a step which is too large and a better point could be found by making a contraction. So go to 4.

Otherwise if

$$F(\mathbf{x}^{(r)}) > F(\mathbf{x}^{(l)})$$

then replace $\mathbf{x}^{(h)}$ by $\mathbf{x}^{(r)}$ and go to 1.

Otherwise, we have obtained the best point yet and it is reasonable to think that a larger step might lead to a better point. An 'expanded' point $\mathbf{x}^{(e)}$ is considered such that:

$$\mathbf{x}^{(e)} - \mathbf{x}^{(g)} = \mu(\mathbf{x}^{(r)} - \mathbf{x}^{(g)}), \qquad \text{with } \mu > 1$$

thus

$$\mathbf{x}^{(e)} = \mathbf{x}^{(g)}(1 - \mu) + \mu\mathbf{x}^{(r)}$$

If

$$F(\mathbf{x}^{(e)}) < F(\mathbf{x}^{(l)})$$

replace $\mathbf{x}^{(h)}$ by $\mathbf{x}^{(e)}$ and go to 1.

Else, replace $\mathbf{x}^{(h)}$ by $\mathbf{x}^{(r)}$ and go to 1. Figure 4.2.5 illustrates the expansion in the simplex.

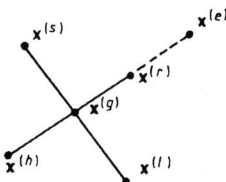

Figure 4.2.5 The expansion in the simplex method.

4. If

$$F(\mathbf{x}^{(r)}) > F(\mathbf{x}^{(h)})$$

a search is made along the line joining $\mathbf{x}^{(h)}$ and $\mathbf{x}^{(g)}$ for a point $\mathbf{x}^{(c)}$ that gives an improvement over $F(\mathbf{x}^{(r)})$.

We make a contraction of the original simplex with

$$\mathbf{x}^{(c)} - \mathbf{x}^{(g)} = \lambda(\mathbf{x}^{(h)} - \mathbf{x}^{(g)}), \qquad \text{with } 0 < \lambda < 1$$

which gives

$$\mathbf{x}^{(c)} = \mathbf{x}^{(g)}(1 - \lambda) + \lambda\mathbf{x}^{(h)}$$

Figure 4.2.6 illustrates this contraction.

Otherwise we consider a point $\mathbf{x}^{(c)}$ on the line joining the points $\mathbf{x}^{(g)}$ and $\mathbf{x}^{(r)}$, then

$$\mathbf{x}^{(c)} = \mathbf{x}^{(g)}(1 - \lambda) + \lambda\mathbf{x}^{(r)}$$

If
$$F(\mathbf{x}^{(c)}) < F(\mathbf{x}^{(h)})$$

replace $\mathbf{x}^{(h)}$ by $\mathbf{x}^{(c)}$ and go to 1.

Figure 4.2.6 illustrates this contraction.

Otherwise the minimum may be enclosed by the present simplex; then the simplex is contracted around the best point by reducing each side by half. That is, each $\mathbf{x}^{(i)}$ is replaced by $(\mathbf{x}^{(i)} + \mathbf{x}^{(l)})/2$, and go to 1

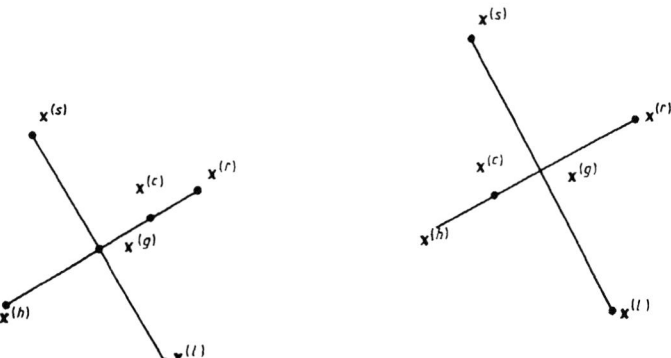

Figure 4.2.6 The contractions in the simplex method.

The choice of the initial simplex is arbitrary. The option S(implex makes a simplex using an initial point $\mathbf{x}^{(1)}$ and the points

$$\mathbf{x}^{(2)} = \mathbf{x}^{(1)} + k\mathbf{e}^{(1)}$$

$$\mathbf{x}^{(3)} = \mathbf{x}^{(1)} + k\mathbf{e}^{(2)}$$

where k is a given step.

Exercise

Find the minimum of BEDP using the option S(implex.

> **4.3   Summary**

The univariate search and the DSC algorithm conform to the pattern for optimisation discussed in Chapter 1. That is, the minimum is found by a sequence of line searches. Each of the one-dimensional minimisations uses a procedure which requires a substantial number of function evaluations. We shall see later that the one-dimensional minimisation is too costly in terms of function evaluations and it is efficient to accept crude estimates of the position of the minimum even though more cycles will then be needed. The practical implications of this will be discussed in Chapter 6.

The Simplex method, though it looks unlike the other methods, does in fact follow the pattern of searching for the minimum along a sequence of directions. As it was presented here, the method in fact uses inexact line searches.

The main advantage of the direct search methods is that the determination of the search direction is not based on the local properties of the function. This makes the methods very suitable for the optimisation of noisy functions, that is, functions which are composed of a small random component superimposed on a general trend.

> Chapter 5

> Gradient methods I

> 5.1 Introduction

In Chapter 3 we discussed methods which search for the minimum by computing trial values of the function. However, we may know more about the function at a given point than just its value. In particular its first partial derivatives may be easy to compute. In this chapter we introduce methods which make use of this information.

The n first partial derivatives of a function of n variables can be assembled as an n-dimensional vector $\mathbf{g}$:

$$
\mathbf{g} = \left[\begin{array}{c} \frac{\partial F}{\partial x_1} \\ \vdots \\ \frac{\partial F}{\partial x_n} \end{array} \right] \tag{5.1.1}
$$

Since the vector $\mathbf{g}$, evaluated at a point x, provides information about the trend in the value of F as x is changed, we might expect it to be useful in determining the best direction of search. This is indeed the case and in this chapter we shall introduce ways of using $\mathbf{g}$ to develop efficient search algorithms.

> 5.2 The first derivative vector

We shall now discuss the most important property of $\mathbf{g}$, which is that it can be interpreted as the direction along which the function F is increasing most rapidly. We shall use a function of two variables to demonstrate this, but the extension to n dimensions is straightforward.

Consider the function F of the two-dimensional vector $\mathbf{x}$ which has first derivatives $\mathbf{g}$ at the point $\mathbf{x}$. Let $\mathbf{d}$ be any two-dimensional vector, and consider the effect of moving a small step s along $\mathbf{d}$ to the point $\mathbf{x} + s\mathbf{d}$.

We have as in equation 2.1.7

$$F(\mathbf{x} + s\mathbf{d}) \approx F(\mathbf{x}) + s\left(d_1\frac{\partial F}{\partial x_1} + d_2\frac{\partial F}{\partial x_2}\right) \qquad (5.2.1)$$

The approximation can be made as accurate as we like by reducing the size of s.

The distance moved is given by

$$\delta l = s\sqrt{d_1^2 + d_2^2} \qquad (5.2.2)$$

Hence, denoting the change in F by δF, we can write:

$$\frac{\delta F}{\delta l} \approx \frac{d_1\frac{\partial F}{\partial x_1} + d_2\frac{\partial F}{\partial x_2}}{\sqrt{d_1^2 + d_2^2}} \qquad (5.2.3)$$

(Although the above expression does not contain s it is still an approximation because it is based on equation 5.2.1).

In terms of the angle θ that the vector $\mathbf{d}$ makes with the x_1 axis as illustrated in Figure 5.2.1, the above expression can be written:

$$\frac{\delta F}{\delta l} \approx \cos(\theta)\frac{\partial F}{\partial x_1} + \sin(\theta)\frac{\partial F}{\partial x_2} \qquad (5.2.4)$$

Let us find the values of θ which give the minimum value of $\delta F/\delta l$. This occurs when the necessary condition (equation 2.1.3) is satisfied:

$$\frac{d}{d\theta}\left(\frac{\delta F}{\delta l}\right) = 0$$

That is, differentiating equation 5.2.4 with respect to θ:

$$-\sin(\theta)\frac{\partial F}{\partial x_1} + \cos(\theta)\frac{\partial F}{\partial x_2} = 0$$

Therefore

$$\tan(\theta) = d_2/d_1 = \frac{\partial F}{\partial x_2}\Big/\frac{\partial F}{\partial x_1}$$

The solutions for this are:

$$\mathbf{d} = \alpha\mathbf{g} \qquad (5.2.5)$$

with α a constant. (Of course, this result is exact only for vanishingly small δl). The solution for $\mathbf{d}$ when α is a negative constant corresponds to a minimum of $\delta F/\delta l$ rather than any other kind of a stationary point and to a maximum when α is a positive constant. This is easy to prove by considering the second derivative condition and this is left as an exercise.

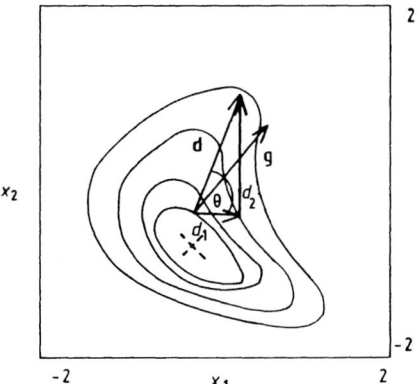

Figure 5.2.1 The direction of vector $\mathbf{d}$.

Equation 5.2.5 shows that for any small step of fixed length δl, the maximum change in F occurs when the step is in the direction of $\mathbf{g}$. Hence $\mathbf{g}$ is the direction of steepest ascent, and $-\mathbf{g}$ the direction of steepest descent.

We have thus shown that the negative of the first derivative vector is the direction of steepest descent.

Another useful property of the vector $\mathbf{g}$ is that it is orthogonal to the tangent plane to F.

Again we shall demonstrate this for the two-dimensional case.

The tangent to the contour of F is the line along which a small step produces no change in F. Thus for a vector $\mathbf{t}$ along the tangent we have from equation 5.2.3:

$$\frac{\delta F}{\delta l} \approx \frac{t_1 \frac{\partial F}{\partial x_1} + t_2 \frac{\partial F}{\partial x_2}}{\sqrt{t_1^2 + t_2^2}} = 0$$

from which we can obtain a condition for the components of **d**:

$$\frac{t_1}{t_2} = -\frac{\partial F/\partial x_2}{\partial F/\partial x_1}, \quad \text{if } \partial F/\partial x_1 \neq 0$$

We have one equation to determine the two components of **t**, therefore, as before, the condition is satisfied if:

$$\mathbf{t} = \begin{bmatrix} -\frac{\partial F}{\partial x_2} \\ \\ \frac{\partial F}{\partial x_1} \end{bmatrix} \tag{5.2.6}$$

To prove that this vector is orthogonal to the gradient vector we take the inner product of **g** and **t**, which gives:

$$\mathbf{g}^t\mathbf{t} = \begin{bmatrix} \frac{\partial F}{\partial x_1} & \frac{\partial F}{\partial x_2} \end{bmatrix} \begin{bmatrix} -\frac{\partial F}{\partial x_2} \\ \\ \frac{\partial F}{\partial x_1} \end{bmatrix} = 0$$

hence the vectors are orthogonal.

In summary, these results extended to the n-dimensional case are:

⊙ The direction of steepest descent on F at any point **x** is the vector whose elements are the negatives of the first partial derivatives of F at **x**; this is the negative of the gradient vector **g**.

• The gradient vector is orthogonal to the tangent plane at **x**.

Example

Let us consider the function BEDP from Chapter 1:

$$F(\mathbf{x}) = 2x_1 + x_2 + x_1^2 - x_2^2 + (x_1 - x_2^2)^2$$

The gradient vector is:

$$\mathbf{g} = \begin{bmatrix} \frac{\partial F}{\partial x_1} \\ \\ \frac{\partial F}{\partial x_2} \end{bmatrix} = \begin{bmatrix} 2 + 4x_1 - 2x_2^2 \\ 1 - 2x_2 - 4(x_1 - x_2^2)x_2 \end{bmatrix} \tag{5.2.7}$$

which at the point:

$$\mathbf{x}^{(1)} = (-1, -1.3)$$

takes the value

$$\mathbf{g}^{(1)} = \begin{bmatrix} -5.38 \\ -10.39 \end{bmatrix}$$

The modulus of $\mathbf{g}^{(1)}$ is $g^{(1)} = 11.7$ and it measures the rate of change of F at $\mathbf{x}^{(1)}$ per unit step along the direction of $\mathbf{g}$.

We wish to show $-\mathbf{g}^{(1)}$ on the contour plot of BEDP. To do this load BEDP with PREPARE and use CONTOUR to plot $\mathbf{x}^{(1)}$. In the stored version of BEDP the vector $-\mathbf{g}$ would lie outside the display window. As we are only interested in the direction of $-\mathbf{g}$ we consider the vector $-\mathbf{g}'$ along $-\mathbf{g}$ such that the vertical component of $-\mathbf{g}' - \mathbf{x}^{(1)}$ is 2, that is:

$$-\mathbf{g}'^{(1)} = \alpha \begin{bmatrix} 5.38 \\ 10.39 \end{bmatrix}$$

and we determine α such that the vertical component of $-\mathbf{g}'$ lies in the window, thus

$$10.39\alpha - 1.3 = 2, \text{ giving } \alpha = 0.3$$

Then

$$-\mathbf{g}' = \begin{bmatrix} 1.61 \\ 3.1 \end{bmatrix}$$

Entering these values using the option **V(ector**, the screen should display $-\mathbf{g}'$ on the contour map and in the top window a plot of the function F along $-\mathbf{g}'$ which shows the fastest decrease of F in the neighbourhood of $\mathbf{x}^{(1)}$. Figure 5.2.2 exhibits this.

The vector

$$\mathbf{t} = \begin{bmatrix} 10.39 \\ -5.38 \end{bmatrix}$$

is tangent to the contour of F at $\mathbf{x}^{(1)}$. To display it with CONTOUR we need to scale it so it fits in the picture window.

We plot the vector $\mathbf{t}' = \alpha'\mathbf{t}$, that is:

$$\mathbf{t}' = \alpha' \begin{bmatrix} 10.39 \\ -5.38 \end{bmatrix}$$

and select α' so the vector remains in the picture window by making the vertical coordinate of $-\mathbf{t}' - \mathbf{x}^{(1)}$ equal to -2, which gives $\alpha' = 0.13$. Thus

$$\mathbf{t}' = \begin{bmatrix} 1.35 \\ -0.7 \end{bmatrix}$$

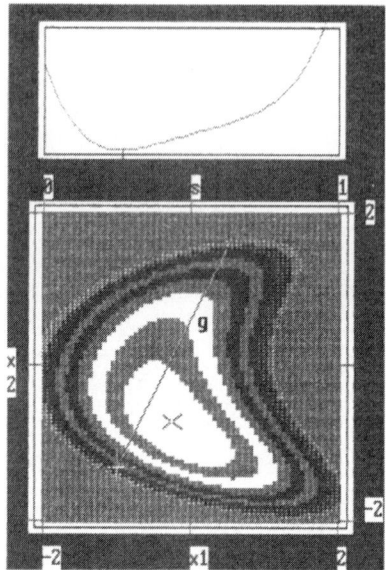

Figure 5.2.2 The direction of steepest descent.

Entering these values the display shows the vectors $\mathbf{t}'$ and $\mathbf{g}'$. Note that the vectors may not look quite at right angles to each other due to the differences in the vertical and horizontal resolutions of the screen. The top window shows the function F along $\mathbf{t}'$, the plot shows the slowest decrease of F in the vicinity of $\mathbf{x}^{(1)}$. Figure 5.2.3 shows this.

Exercise

Display the gradient vector and the tangent to the contour of the function BEDP at the point $\mathbf{x}^{(2)} = (1, 0)$. Calculate the modulus of the gradient at $\mathbf{x}^{(2)}$; you should find that the modulus is much smaller than at $\mathbf{x}^{(1)}$ where the contours of F are closer together, indicating a steeper change in F.

We shall now develop an algorithm based on searches along the direction of steepest descent.

> 5.3   The steepest descent algorithm

The steepest descent algorithm is as follows:

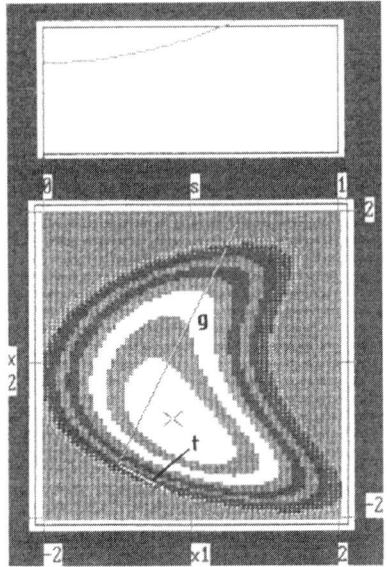

Figure 5.2.3 The direction of least change.

1. Select some arbitrary starting point $\mathbf{x}^{(0)}$. Set $k = 0$.

2. Calculate $\mathbf{g}^{(k)}$. Stop if

$$\|\mathbf{g}^{(k)}\| < \epsilon_1$$

3. Set the search direction to

$$-\mathbf{g}^{(k)}$$

4. Compute the new point $\mathbf{x}^{(k+1)}$ by searching for the minimum along this direction.

5. Calculate $\mathbf{g}^{(k+1)}$. Stop if

$$\|\mathbf{x}^{(k+1)} - \mathbf{x}^{(k)}\| < \epsilon_2$$

or

$$\|\mathbf{g}^{(k+1)}\| < \epsilon_1$$

where ϵ_1 and ϵ_2 are small positive numbers.
Otherwise set $k = k + 1$ and repeat 3.

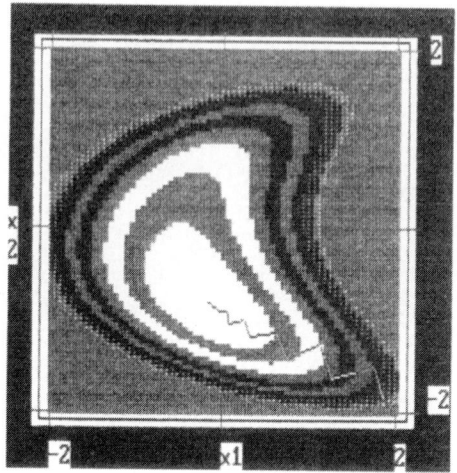

Figure 5.3.1 The steepest descent algorithm.

Example

Let us find the minimum of BEDP using the steepest descent method. For this we use the **S(earch S(teep** option in the program STEEP. Figure 5.3.1 illustrates the algorithm with starting point at (2, 2). Notice that in the picture each direction is orthogonal to the one before.

To see why this should be so, consider any search direction **d**. A minimum along **d** is by definition a point where the derivative of F along **d** is zero. That is, at that point **d** is the direction along which $\partial F / \partial l = 0$. It is therefore a tangent to the contour there. Since this is true for any direction **d** it must also be true for steepest descent directions. Hence in the steepest descent method each search direction begins orthogonal to a tangent to a contour and ends as a tangent to another contour. Thus, each direction is orthogonal to the one before.

Exercises

Use the option **S(teep** for the method of steepest descent in the program STEEP:

- to find the minimum of BEDP with starting point at $(2, -2)$

- to try to find the minimum of Rosenbrock's function:

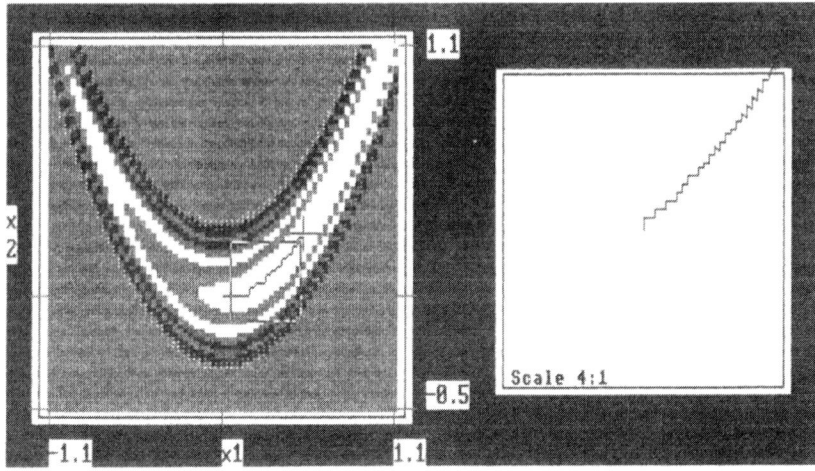

Figure 5.3.2 Inefficiency in the steepest descent algorithm.

$$F(\mathbf{x}) = 100(x_2 - x_1^2)^2 + (1 - x_1)^2$$

with starting point at $(-0.4, 0.2)$.

Steepest descent is a reliable algorithm which can be shown to converge to a local minimum on any well behaved function, provided exact arithmetic is used. In this it resembles the algorithms of Chapter 3. However, like the univariate search it can be inefficient; you should have experienced this characteristic in the above exercise. The inefficiency is illustrated in Figure 5.3.2 which shows the line searches following a zig-zag pattern towards the minimum.

It would be interesting if the inefficiency of the algorithm could be overcome by some simple change that would modify the steepest descent algorithm and cause it to adjust the search directions to the geometry of the function, as the DSC algorithm did for univariate search. It turns out that when search directions are formed by adding to each gradient vector a proportion of the previous search direction a much more efficient algorithm results. However before we derive the modified algorithm we must discuss the matrix formed by the second derivatives of F and introduce the *quadratic function of n variables*.

> 5.4   The second derivative matrix

We can express any smooth continuous function F near any point $\mathbf{x}^*$ by its Taylor's expansion:

$$F(\mathbf{x}) \approx F(\mathbf{x}^*) + (\mathbf{x} - \mathbf{x}^*)^t \mathbf{g}^* + \frac{1}{2}(\mathbf{x} - \mathbf{x}^*)^t \mathbf{G}^*(\mathbf{x} - \mathbf{x}^*)$$

where $\mathbf{g}^*$ is the gradient vector evaluated at $\mathbf{x}^*$. The matrix $\mathbf{G}^*$ whose components are

$$G_{ij}^* = \frac{\partial^2 F}{\partial x_i \partial x_j} \tag{5.4.1}$$

with the partial derivatives evaluated at $\mathbf{x}^*$, is called the **hessian matrix**.

For $\mathbf{x}^*$ to be a stationary point we must have:

$$\mathbf{g}^* = 0$$

Thus for $\mathbf{x}$ in the neighbourhood of $\mathbf{x}^*$ we have:

$$F(\mathbf{x}) \approx F(\mathbf{x}^*) + \frac{1}{2}(\mathbf{x} - \mathbf{x}^*)^t \mathbf{G}^*(\mathbf{x} - \mathbf{x}^*)$$

A sufficient condition for $\mathbf{x}^*$ to be a minimum is

$$F(\mathbf{x}^*) < F(\mathbf{x})$$

for

$$|\mathbf{x} - \mathbf{x}^*| < \epsilon$$

Thus

$$(\mathbf{x} - \mathbf{x}^*)^t \mathbf{G}^*(\mathbf{x} - \mathbf{x}^*) > 0 \tag{5.4.2}$$

for all

$$\mathbf{x} \neq \mathbf{x}^*$$

When $\mathbf{G}^*$ satisfies the above condition it is said to be **positive definite**.
Similarly for $\mathbf{x}^*$ to be a maximum we must have:

$$(\mathbf{x} - \mathbf{x}^*)^t \mathbf{G}^*(\mathbf{x} - \mathbf{x}^*) < 0 \tag{5.4.3}$$

for all

$$\mathbf{x} \neq \mathbf{x}^*$$

The matrix G^* is said to be **negative definite** when it satisfies the above condition.

When both conditions 5.4.2 and 5.4.3 can be satisfied for some x in the vicinity of x^* then x^* is a **saddle point** and G^* is said to be indefinite.

In general, any matrix M *is said to be positive definite* if and only if it satisfies the condition:

$$y^t M y > 0, \quad \text{for all compatible non-zero vectors } y. \qquad (5.4.4)$$

It is not in general easy to check the definiteness of a matrix using equation 5.4.4. However, one possible test for definiteness is based on the fact that the hessian is a symmetric matrix and on a property of symmetric matrices. We shall simply quote and apply this property. More details can be found in books on matrix methods, see for example reference [5].

A symmetric matrix G is positive definite if and only if all its eigenvalues are positive.

The eigenvalues of G are the solutions of the equation:

$$\det(G - \lambda I) = 0 \qquad (5.4.5)$$

For 2 x 2 matrices it is straightforward to solve equation 5.4.5, although for larger matrices the method is not practical. There are more generally applicable methods but they are beyond the scope of this book; the reader is referred to reference [5].

The definiteness of the hessian matrix can be used to classify the stationary points of a function F as follows:

F has a stationary point at x if

$$g(x) = 0$$

1. If G is positive definite (all $\lambda_i > 0$), x is a minimum.

2. If G is negative definite (all $\lambda_i < 0$), x is a maximum.

3. If G is indefinite (λ_i of mixed signs), x is a saddle point.

4. If any of the λ_i is zero, G is said to be singular and this test is not applicable.

Example

Let us classify the stationary points of the function

$$F(\mathbf{x}) = 2x_1^2 + x_1 x_2^2 + x_2^2$$

The stationary points are the solutions of:

$$\frac{\partial F}{\partial x_1} = 4x_1 + x_2^2 = 0$$

$$\frac{\partial F}{\partial x_2} = 2x_1 x_2 + 2x_2 = 0$$

which give:

$$\mathbf{x}^{(1)} = (0,0), \quad \mathbf{x}^{(2)} = (-1,2) \quad \mathbf{x}^{(3)} = (-1,-2).$$

The hessian matrix is

$$\mathbf{G} = \begin{bmatrix} 4 & 2x_2 \\ 2x_2 & 2x_1 + 2 \end{bmatrix}$$

Thus:

$$\mathbf{G}^{(1)} = \begin{bmatrix} 4 & 0 \\ 0 & 2 \end{bmatrix}$$

The eigenvalues are the solution of

$$(4 - \lambda)(2 - \lambda) = 0$$

which gives

$$\lambda_1 = 4, \quad \lambda_2 = 2$$

Thus $\mathbf{x}^{(1)}$ corresponds to a minimum. Similarly

$$\mathbf{G}^{(2)} = \begin{bmatrix} 4 & 4 \\ 4 & 0 \end{bmatrix}$$

has eigenvalues

$$\lambda_1 = 2 + \sqrt{20}, \quad \lambda_2 = 2 - \sqrt{20}$$

Thus $\mathbf{x}^{(2)}$ corresponds to a saddle point. Finally

$$\mathbf{G}^{(3)} = \begin{bmatrix} 4 & -4 \\ -4 & 0 \end{bmatrix}$$

has the same eigenvalues as $\mathbf{G}^{(2)}$ and therefore $\mathbf{x}^{(3)}$ corresponds to a saddle point.

Exercise

Show that for the function of two variables $F(x_1, x_2)$ to have a minimum at the point $\mathbf{x}$ all the second derivatives $\partial^2 F / \partial x_i{}^2$, $i = 1, 2$, must be positive but the mixed second derivative $\partial^2 F / \partial x_1 \partial x_2$ need not be.

> 5.5 Quadratic functions

A particular type of n-dimensional function which is of interest is the quadratic function $Q(\mathbf{y})$ defined as follows:

$$Q(\mathbf{y}) = a + \mathbf{y}^t \mathbf{b} + \frac{1}{2} \mathbf{y}^t \mathbf{H} \mathbf{y} \qquad (5.5.1)$$

where a is a scalar constant, $\mathbf{b}$ is a constant vector and $\mathbf{H}$ is a constant symmetric matrix.

The quadratic function formed by the first three terms of the Taylor's expansion of a function F about any point can be used as its approximation in the neighbourhood of that point.

Therefore it is reasonable to expect that optimisation methods which are efficient for quadratic functions should also perform well for more general functions. In particular near the optimum, where the minimum of the quadratic approximation is close to the true minimum, the efficiency of such methods should be high.

The approach of using a quadratic approximation to optimise a function of n variables is similar to the one used in linear searches where the optimum of a function of one variable is determined by a succession of optimisations of its quadratic approximation.

Example

Consider the quadratic approximation to the function BEDP. We have already calculated its gradient vector in equation 5.2.7. The hessian matrix is:

$$\mathbf{G} = \begin{bmatrix} 4 & -4x_2 \\ -4x_2 & -2 - 4(x_1 - 3x_2^2) \end{bmatrix}$$

Let us approximate the function about the point $\mathbf{x}^{(1)} = (0, -0.7)$. The gradient and the hessian at this point are:

$$\mathbf{g}^{(1)} = \left[\begin{array}{c} 1.02 \\ 1.028 \end{array} \right], \quad \mathbf{H}^{(1)} = \left[\begin{array}{cc} 4 & 2.8 \\ 2.8 & 3.88 \end{array} \right]$$

The form of the quadratic approximation about $\mathbf{x}$ is:

$$Q(\mathbf{x}) = -0.9499 + 1.02x_1 + 1.028x_2 + 2x_1^2 + \frac{3.88}{2}x_2^2 + 3.88x_1x_2$$

The contours of this function can be obtained using the program PREPARE. By storing the contours on a disc, the screen can be made to change from showing the contours of BEDP to showing the contours of Q. By loading in turn the contours of BEDP and Q one can observe that only in the vicinity of $\mathbf{x}^{(1)}$ is the quadratic function Q a good approximation to BEDP

This illustrates that the quadratic approximation at a point is a useful guide to the position of the minimum so long as the point is itself sufficiently close to the minimum.

Exercise

Use the program PREPARE to study the contours of the following quadratic functions:

-

$$Q(\mathbf{x}) = 1 + x_1^2 + x_2^2$$

that is

$$Q(\mathbf{x}) = 1 + \left[\begin{array}{cc} x_1 & x_2 \end{array} \right] \left[\begin{array}{cc} 1 & 0 \\ 0 & 1 \end{array} \right] \left[\begin{array}{c} x_1 \\ x_2 \end{array} \right]$$

The contours are circles centred at the origin.

-

$$Q(\mathbf{x}) = 1 + x_1 + 2x_2 + x_1^2 + x_2^2$$

that is

$$Q(\mathbf{x}) = 1 + \left[\begin{array}{cc} x_1 & x_2 \end{array} \right] \left[\begin{array}{c} 1 \\ 2 \end{array} \right] + \left[\begin{array}{cc} x_1 & x_2 \end{array} \right] \left[\begin{array}{cc} 1 & 0 \\ 0 & 1 \end{array} \right] \left[\begin{array}{c} x_1 \\ x_2 \end{array} \right]$$

A linear term added to the previous function moves the position of the minimum.

●

$$Q(\mathbf{x}) = 1 + x_1 + 2x_2 + 10x_1^2 + x_2^2,$$

that is

$$Q(\mathbf{x}) = 1 + \begin{bmatrix} x_1 & x_2 \end{bmatrix} \begin{bmatrix} 1 \\ 2 \end{bmatrix} + \frac{1}{2} \begin{bmatrix} x_1 & x_2 \end{bmatrix} \begin{bmatrix} 20 & 0 \\ 0 & 2 \end{bmatrix} \begin{bmatrix} x_1 \\ x_2 \end{bmatrix}$$

When **H** is a diagonal matrix with non-equal elements, the contours are ellipses parallel to the coordinate axes.

●

$$Q(\mathbf{x}) = 1 + x_1 + 2x_2 + 10x_1^2 + x_2^2 + x_1 x_2$$

that is

$$Q(\mathbf{x}) = 1 + \begin{bmatrix} x_1 & x_2 \end{bmatrix} \begin{bmatrix} 1 \\ 2 \end{bmatrix} + \frac{1}{2} \begin{bmatrix} x_1 & x_2 \end{bmatrix} \begin{bmatrix} 20 & 1 \\ 1 & 2 \end{bmatrix} \begin{bmatrix} x_1 \\ x_2 \end{bmatrix}$$

When non-diagonal elements are added to **H**, the contours are rotated.

Finally, before we discuss the next algorithm, we need to introduce the concept of *conjugate vectors.* Non-zero vectors **u** and **v** are said to be conjugate with respect to a non-singular matrix **H** if and only if

$$\mathbf{u}^t \mathbf{H} \mathbf{v} = 0$$

> ## 5.6 The conjugate gradient algorithm

We shall first show that the minimum of any quadratic function of n variables can be found by searching along at most n independent directions which are mutually conjugate with respect to their hessian matrix.

For simplicity we shall prove this first for functions of two variables.

Consider the quadratic function given in equation 5.5.1 where the vector **y** has components which denote coordinates on a set of orthogonal axes. Let us express Q in terms of coordinate axes which are conjugate with respect to the hessian matrix **H**. Let $\mathbf{c}_1$ and $\mathbf{c}_2$ be the two conjugate vectors referred to the orthogonal system.

A point with coordinates y_1 and y_2 in the orthogonal system has coordinates z_1 and z_2 in the conjugate system where:

$$\mathbf{y} = z_1 \mathbf{c}_1 + z_2 \mathbf{c}_2 \qquad (5.6.1)$$

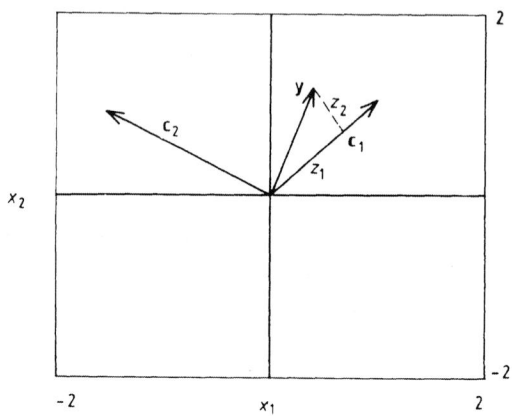

Figure 5.6.1 Conjugate axes.

Figure 5.6.1 illustrates this.

Substituting for **y** in equation 5.5.1 gives:

$$Q(\mathbf{z}) = a + (z_1\mathbf{c}_1 + z_2\mathbf{c}_2)^t\mathbf{b} + \frac{1}{2}(z_1\mathbf{c}_1 + z_2\mathbf{c}_2)^t\mathbf{H}(z_1\mathbf{c}_1 + z_2\mathbf{c}_2)$$

and multiplying out the last term:

$$Q(\mathbf{z}) = a + (z_1\mathbf{c}_1 + z_2\mathbf{c}_2)^t\mathbf{b} + \frac{1}{2}(z_1^2\mathbf{c}_1^t\mathbf{H}\mathbf{c}_1 + 2z_1z_2\mathbf{c}_1^t\mathbf{H}\mathbf{c}_2 + z_2^2\mathbf{c}_2^t\mathbf{H}\mathbf{c}_2)$$

But for conjugate vectors:

$$\mathbf{c}_1^t\mathbf{H}\mathbf{c}_2 = 0$$

hence

$$Q(\mathbf{z}) = a + (z_1\mathbf{c}_1^t\mathbf{b} + \frac{1}{2}z_1^2\mathbf{c}_1^t\mathbf{H}\mathbf{c}_1) + (z_2\mathbf{c}_2^t\mathbf{b} + \frac{1}{2}z_2^2\mathbf{c}_2^t\mathbf{H}\mathbf{c}_2).$$

Note that all the quantities in the right hand side of the above expression are constants except z_1 and z_2.

To obtain the minimum of Q we can minimise the first bracket as a function only of z_1 without affecting the other terms. Similarly for the second bracket. In other words, we can minimise Q by independent line searches along the two directions $\mathbf{c}_1$ and $\mathbf{c}_2$.

This is what we set out to prove, and the result can be readily generalised to n dimensions.

Example

Consider the quadratic function:

$$Q(\mathbf{x}) = 0 + \mathbf{y}^t \begin{bmatrix} 1 \\ 1 \end{bmatrix} + \frac{1}{2}\mathbf{y}^t \begin{bmatrix} 1 & \frac{1}{2} \\ \frac{1}{2} & 2 \end{bmatrix} \mathbf{y} \qquad (5.6.2)$$

The **b** vector is

$$\mathbf{b} = \begin{bmatrix} 1 \\ 1 \end{bmatrix}$$

and the hessian matrix is

$$\mathbf{H} = \begin{bmatrix} 1 & \frac{1}{2} \\ \frac{1}{2} & 2 \end{bmatrix} \qquad (5.6.3)$$

It is easy to check that the vectors

$$\mathbf{c}_1 = \begin{bmatrix} 1 \\ 1 \end{bmatrix} \text{ and } \mathbf{c}_2 = \begin{bmatrix} -5/3 \\ 1 \end{bmatrix} \qquad (5.6.4)$$

are conjugate with respect to the hessian matrix given in equation 5.6.3. (Verify this for yourself). We have not yet discussed how to generate vectors with such a property. We shall do this later; for now we assume that the conjugate vectors $\mathbf{c}_1$ and $\mathbf{c}_2$ are given.

The vectors are shown in Figure 5.6.1 and they can be considered as forming a coordinate system. To express the function Q in terms of the c-system we need to use the transformation given in equation 5.6.1. Using this transformation the different terms take values:

$$\mathbf{c}_1^t \mathbf{b} = \begin{bmatrix} 1 & 1 \end{bmatrix} \begin{bmatrix} 1 \\ 1 \end{bmatrix} = 2$$

$$\mathbf{c}_2^t \mathbf{b} = \begin{bmatrix} -5/3 & 1 \end{bmatrix} \begin{bmatrix} 1 \\ 1 \end{bmatrix} = -2/3$$

$$\mathbf{c}_1^t \mathbf{H} \mathbf{c}_1 = \begin{bmatrix} 1 & 1 \end{bmatrix} \begin{bmatrix} 1 & \frac{1}{2} \\ \frac{1}{2} & 2 \end{bmatrix} \begin{bmatrix} 1 \\ 1 \end{bmatrix} = 4$$

$$\mathbf{c}_2^t \mathbf{H} \mathbf{c}_2 = \begin{bmatrix} -5/3 & 1 \end{bmatrix} \begin{bmatrix} 1 & \frac{1}{2} \\ \frac{1}{2} & 2 \end{bmatrix} \begin{bmatrix} -5/3 \\ 1 \end{bmatrix} = 28/9.$$

Hence
$$Q(\mathbf{z}) = (2z_1 + 2z_1^2) + (-2/3z_2 + 14/9z_2^2)$$

Minimising each term in turn we find, by minimising with respect to z_1, i.e. along the direction given by $\mathbf{c}_1$:

$$\min(2z_1 + 2z_1^2) = -\frac{1}{2}, \quad \text{which occurs when } z_1 = -\frac{1}{2}$$

and, by minimising with respect to z_2, i.e. along the direction given by $\mathbf{c}_2$:

$$\min(-2/3z_2 + 14/9z_2^2) = -1/14, \quad \text{which occurs when } z_1 = 3/14.$$

Transforming back into the y-system, the minimum of Q is at:

$$\mathbf{y}^{(min)} = -\frac{1}{2}\begin{bmatrix} 1 \\ 1 \end{bmatrix} + 3/14\begin{bmatrix} -5/3 \\ 1 \end{bmatrix}$$

then

$$\mathbf{y}^{(min)} = \begin{bmatrix} -6/7 \\ -2/7 \end{bmatrix}$$

Note that the c-system is not formed by unit coordinate vectors, therefore the distance from the origin to the projection of $\mathbf{y}^{(min)}$ on $\mathbf{c}_1$ is:

$$\frac{1}{2}\sqrt{1+1} = 0.7071$$

and similarly for the other conjugate direction:

$$3/14\sqrt{(5/3)^2 + 1} = 0.417$$

Exercise

Using the program CONTOUR check that the minimum of Q from equation 5.6.2 can be obtained by two independent line searches along the directions $\mathbf{c}_1$ and $\mathbf{c}_2$ given by the vectors of equations 5.6.4. Check that the procedure works for any starting point.

Summarising, an n-dimensional quadratic function can be minimised in n steps given n mutually conjugate directions with respect to the hessian matrix.

However, there remains the problem of determining the conjugate directions. There are many effective methods which generate such directions one at a time, as they are needed. One of these methods requires only a small modification to the steepest descent algorithm; it is known as the *conjugate gradient*, or *Fletcher–Reeves method.*

The conjugate gradient method for a general function is as follows:

1. Select some arbitrary starting point $x^{(0)}$. Set ϵ_1 and ϵ_2 as two small positive numbers. Set $k = 0$.

2. Calculate $g^{(k)}$. Stop if

$$|g^{(k)}| < \epsilon_1$$

Otherwise set the initial search direction

$$p^{(0)} = -g^{(0)}$$

3. Compute the new point $x^{(k+1)}$ by searching for the minimum along this direction.

4. Stop if
$$|x^{(k+1)} - x^{(k)}| < \epsilon_2$$

or

$$|g^{(k+1)}| < \epsilon_1$$

Otherwise:

5. Compute the new search direction

$$p^{(k+1)} = -g^{(k+1)} + \beta^{(k+1)}p^{(k)}$$

where

$$\beta^{(k+1)} = \frac{(g^{(k+1)})^t g^{(k+1)}}{(g^{(k)})^t g^{(k)}}$$

Set $k = k + 1$ and repeat 4.

This may be seen as a modification to the steepest descent algorithm, made by forming a new search direction, by adding a proportion of the previous search direction to the new gradient.

The usefulness of the method lies in the fact that the conjugate directions for a quadratic function are determined without calculating the hessian matrix and that storage is required for only two vectors.

We shall now show that the above algorithm does generate conjugate directions on two-dimensional quadratic functions.

We have

$$Q(\mathbf{x}) = a + \mathbf{b}^t\mathbf{x} + \frac{1}{2}\mathbf{x}^t\mathbf{H}\mathbf{x}$$

and

$$\mathbf{p}^{(0)} = -\mathbf{g}^{(0)} = -\left(\mathbf{b} + \mathbf{H}\mathbf{x}^{(0)}\right) \tag{5.6.5}$$

The new point is at

$$\mathbf{x}^{(1)} = \mathbf{x}^{(0)} + s^{(0)}\mathbf{p}^{(0)} \tag{5.6.6}$$

and the new direction is

$$\mathbf{p}^{(1)} = -\mathbf{g}^{(1)} - \beta^{(1)}\mathbf{g}^{(0)}$$

Hence

$$\mathbf{p}^{(1)t}\mathbf{H}\mathbf{p}^{(0)} = \left(-\mathbf{g}^{(1)} - \beta^{(1)}\mathbf{g}^{(0)}\right)^t \mathbf{H}\mathbf{p}^{(0)}$$

But, from equation 5.6.6:

$$\mathbf{p}^{(0)} = \frac{\left(\mathbf{x}^{(1)} - \mathbf{x}^{(0)}\right)}{s^{(0)}}$$

Hence

$$\mathbf{p}^{(1)t}\mathbf{H}\mathbf{p}^{(0)} = \frac{1}{s^{(0)}}\left(-\mathbf{g}^{(1)} - \beta^{(1)}\mathbf{g}^{(0)}\right)^t \mathbf{H}\left(\mathbf{x}^{(1)} - \mathbf{x}^{(0)}\right) \tag{5.6.7}$$

Substituting

$$\mathbf{H}\left(\mathbf{x}^{(1)} - \mathbf{x}^{(0)}\right)^t = \mathbf{g}^{(1)} - \mathbf{g}^{(0)}$$

into the right hand side of equation 5.6.7 gives:

$$\frac{1}{s^{(0)}}\left(-\mathbf{g}^{(1)} - \beta^{(1)}\mathbf{g}^{(0)}\right)\left(-\mathbf{g}^{(1)} - \mathbf{g}^{(0)}\right).$$

Multiplying the expression out and recalling that because $\mathbf{x}^{(1)}$ is the position of the minimum of Q along $\mathbf{p}^{(0)}$, then:

$$\mathbf{g}^{(1)t}\mathbf{g}^{(0)} = 0$$

We obtain:
$$p^{(1)t}Hp^{(0)} = 0.$$

Thus the search directions p are mutually conjugate and hence the algorithm generates conjugate search directions in two dimensions.

This proof can be seen to depend on two conditions:

• initial search direction is steepest descent;

• line searches are exact.

We emphasise these two points, because as we shall see in the following chapter, there are other methods which mantain their efficiency without such restrictive conditions.

The proof refers to quadratic functions. When the method is applied to a general function F the hessian is not constant, so conjugacy is not maintained and it may take more than n steps to approach the solution. It is therefore usual to restart the method after every n steps by searching along the steepest descent direction in order to satisfy the first condition above.

To prove the above results for n-dimensional functions is more involved and the reader is referred to [9].

Exercise

Use the method of conjugate directions implemented in the program STEEP to find the minimum of Rosenbrock's function starting at the point $(-0.4, 0.2)$.

> 5.7   Summary

This chapter discussed the effect of the geometry of the function on the problem of finding good search directions. It was found that useful search directions can be constructed using a knowledge of the gradient vector. The steepest descent algorithm is the simpler of the two methods discussed; it is reliable but can be very inefficient.

To overcome this drawback, the conjugate gradient method was introduced. This involves a small modification to the steepest descent method and thus retains its simplicity and low requirement for computer memory. However, it requires very accurate line searches, which are expensive in terms of number of function evaluations. In the next chapter we shall discuss methods which can use less exact line searches and are in general more efficient, although they need more computer memory.

> Chapter 6

> Gradient methods II

> 6.1 Introduction

In the previous chapter we discussed the method of conjugate gradients, in which the effect of the second derivatives of the function was implicitly incorporated in the determination of the search directions. In this chapter we develop methods which take account of these derivatives but the calculations involve either an explicit calculation of the hessian or approximations to it.

> 6.2 Newton's method

The simplest of the methods is Newton's method. To find $\mathbf{x}^{(*)}$, the position of the minimum of a function F, we consider a trial point $\mathbf{x}^{(0)}$ and calculate the quadratic approximation to F about this point:

$$
\begin{aligned}
F(\mathbf{x}) \quad \approx Q(\mathbf{x}) = \quad & F(\mathbf{x}^{(0)}) + (\mathbf{x} - \mathbf{x}^{(0)})^t \mathbf{g}^{(0)} \\
& + \frac{1}{2}(\mathbf{x} - \mathbf{x}^{(0)})^t \mathbf{G}^{(0)}(\mathbf{x} - \mathbf{x}^{(0)}) \\
= \quad & F(\mathbf{x}^{(0)}) + \sum_{i=1}^{n} g_i^{(0)}(x_i - x_i^{(0)}) \\
& + \frac{1}{2}\sum_{i}\sum_{j} G_{ij}(x_i - x_i^{(0)})(x_j - x_j^{(0)})
\end{aligned}
$$

It is reasonable to assume that the point at which Q has its minimum would be a better approximation than $\mathbf{x}^{(0)}$ to the position of the mini-

83

mum of F, provided that $\mathbf{x}^{(0)}$ is sufficiently close to $\mathbf{x}^{(*)}$. At the minimum of Q its gradient vector is zero, that is:

$$
\begin{bmatrix} \partial Q/\partial x_1 \\ \vdots \\ \partial Q/\partial x_n \end{bmatrix} = \begin{bmatrix} (\partial F/\partial x_1)^{(0)} + \sum_{j=1}^{n} \left(\frac{\partial^2 F}{\partial x_1 \partial x_j}\right)^{(0)} (x_j - x_j^{(0)}) \\ \vdots \\ (\partial F/\partial x_n)^{(0)} + \sum_{j=1}^{n} \left(\frac{\partial^2 F}{\partial x_n \partial x_j}\right)^{(0)} (x_j - x_j^{(0)}) \end{bmatrix} = 0
$$

Let $\mathbf{x}^{(1)}$ be the point at which the above equation is satisfied, then we can write it as:

$$
\mathbf{g}^{(0)} + \mathbf{G}^{(0)}(\mathbf{x}^{(1)} - \mathbf{x}^{(0)}) = 0 \tag{6.2.1}
$$

Calling

$$
\mathbf{d}^{(0)} = \mathbf{x}^{(1)} - \mathbf{x}^{(0)}
$$

then the position of the stationary point of Q is:

$$
\mathbf{x}^{(1)} = \mathbf{x}^{(0)} + \mathbf{d}^{(0)}
$$

and $\mathbf{d}^{(0)}$ satisfies the system of linear equations

$$
\mathbf{G}^{(0)}\mathbf{d}^{(0)} = -\mathbf{g}^{(0)} \tag{6.2.2}
$$

whose solution can be formally written:

$$
\mathbf{d}^{(0)} = -(\mathbf{G}^{(0)})^{-1}\mathbf{g}^{(0)} \tag{6.2.3}
$$

Newton's method considers $\mathbf{x}^{(1)}$ as a new approximation to the stationary point of F; if further improvements are required the process must be repeated starting at $\mathbf{x}^{(1)}$.

We now summarise Newton's algorithm:

1. Select some arbitrary starting point $\mathbf{x}^{(0)}$. Set $k = 0$.

2. Calculate $\mathbf{g}^{(k)}$. Stop if $||\mathbf{g}^{(k)}|| < \epsilon_1$. Otherwise calculate $\mathbf{G}^{(k)}$.

3. Solve for $\mathbf{d}^{(k)}$ the system of equations:

$$
\mathbf{G}^{(k)}\mathbf{d}^{(k)} = -\mathbf{g}^{(k)}
$$

Stop if

$$
||\mathbf{d}^{(k)}|| < \epsilon_2
$$

k	$\mathbf{x}^{(k)}$	$F^{(k)}$	$\mathbf{g}^{(k)}$	$\mathbf{G}^{(k)}$
0	$\begin{bmatrix} 1.0 \\ 1.0 \end{bmatrix}$	4.367	$\begin{bmatrix} 5.437 \\ 1.649 \end{bmatrix}$	$\begin{bmatrix} 16.31 & 0 \\ 0 & 3.297 \end{bmatrix}$
1	$\begin{bmatrix} 0.6667 \\ 0.5000 \end{bmatrix}$	2.693	$\begin{bmatrix} 2.079 \\ 0.567 \end{bmatrix}$	$\begin{bmatrix} 5.892 & 0 \\ 0 & 1.416 \end{bmatrix}$
2	$\begin{bmatrix} 0.3137 \\ 0.1000 \end{bmatrix}$	2.108	$\begin{bmatrix} 0.6923 \\ 0.1005 \end{bmatrix}$	$\begin{bmatrix} 2.641 & 0 \\ 0 & 1.015 \end{bmatrix}$
3	$\begin{bmatrix} 0.0516 \\ 0.0010 \end{bmatrix}$	2.003	$\begin{bmatrix} 0.1035 \\ 0.0010 \end{bmatrix}$	$\begin{bmatrix} 2.016 & 0 \\ 0 & 1 \end{bmatrix}$
4	$\begin{bmatrix} 0.0000 \\ 0.0000 \end{bmatrix}$	2.000	$\begin{bmatrix} 0.0005 \\ 0.0000 \end{bmatrix}$	$\begin{bmatrix} 2.00 & 0 \\ 0 & 1 \end{bmatrix}$

Table 6.2.1 Iterations for Newton's method.

4. Determine
$$\mathbf{x}^{(k+1)} = \mathbf{x}^{(k)} + \mathbf{d}^{(k)}$$
Set $k = k + 1$ and repeat from 2.

Example

Let us use Newton's method to find the minimum of

$$F(\mathbf{x}) = \exp(x_1^2) + \exp(0.5x_2^2)$$

It is easy to show that the minimum occurs at the origin. Table 6.2.1 displays the iterations of the method.

Exercise

Repeat the calculations of the above example using the option **N(ewton**

From the table of iterations, it is apparent that after the second iteration the method seems at least to double the number of exact significant figures of the elements of $\mathbf{x}^{(k)}$. This behaviour is an example of quadratic convergence.

A method is said to have **quadratic convergence** if for sufficiently large i:

$$||\mathbf{x}^{(i)} - \mathbf{x}^{(*)}|| \to K||\mathbf{x}^{(i-1)} - \mathbf{x}^{(*)}||^2 \to 0$$

where K is a positive constant.

It can be proved that when the iterations are sufficiently close to the minimum Newton's method has quadratic convergence. A proof is given in reference [10]. Like all proofs of rate of convergence, it makes many assumptions not only about the nature of the function $F(\mathbf{x})$, but also about the behaviour of the algorithm outside the immediate vicinity of $\mathbf{x}^{(*)}$. However, it does give an indication of the final rate of convergence; in effect, once the estimate $\mathbf{x}^{(i)}$ is sufficiently close to a local optimum, the number of correct significant figures in each component of $\mathbf{x}^{(i)}$ should about double on every iteration as illustrated in Table 6.2.1.

Though Newton's method is fast to converge when it is applied sufficiently close to the minimum, it has the serious shortcoming that it may generate points which move away from the position of the optimum as in the following example.

Example

Find the position of the minimum of Rosenbrock's function with starting point at $(0, 0)$:

$$F(\mathbf{x}) = 100(x_2 - x_1^2)^2 + (1 - x_1)^2$$

The gradient vector is:

$$\mathbf{g(x)} = \left[\begin{array}{c} -400x_1(x_2 - x_1^2) - 2(1 - x_1) \\ 200(x_2 - x_1^2) \end{array} \right]$$

The hessian matrix is:

$$\mathbf{G(x)} = \left[\begin{array}{cc} 1200x_1^2 - 400x_2 + 2 & -400x_1 \\ -400x_1 & 200 \end{array} \right]$$

To determine the search direction we need to solve the system of equations:

$$\left[\begin{array}{cc} 2 & 0 \\ 0 & 200 \end{array} \right] \left[\begin{array}{c} d_1^{(0)} \\ d_2^{(0)} \end{array} \right] = \left[\begin{array}{c} 2 \\ 0 \end{array} \right]$$

from which we obtain:

$$\mathbf{d}^{(0)} = \left[\begin{array}{c} 1 \\ 0 \end{array} \right].$$

Thus:

$$\mathbf{x}^{(1)} = \begin{bmatrix} 0 \\ 0 \end{bmatrix} + \begin{bmatrix} 1 \\ 0 \end{bmatrix} = \begin{bmatrix} 1 \\ 0 \end{bmatrix}$$

We observe that $F(\mathbf{x}^{(0)}) = 1$ and $F(\mathbf{x}^{(1)}) = 100$. We have moved to a point with a higher value for the function! Although this does not necessarily lead to divergence, it is a dangerous property because divergence is possible in such conditions.

To avoid such problems the algorithm is modified so that the new iteration point $\mathbf{x}^{(i+1)}$ is determined as the position of the minimum of F along the search direction $\mathbf{d}^{(i)}$, that is

$$\mathbf{x}^{(i+1)} = \mathbf{x}^{(i)} + s^{(i)}\mathbf{d}^{(i)}$$

where $s^{(i)}$ is the value of s which minimises $F(\mathbf{x}^{(i)} + s\mathbf{d}^{(i)})$.

Let us see graphically the effect of this modification in our example. To do this, display on the computer screen the contours of the function and using the program CONTOUR show the point with coordinates $(0,0)$ and from it the vector $\mathbf{d}^{(0)}$. The upper window should exhibit the values of F along the line drawn on the contour plot with the minimum along the line at a distance of 0.1562 from $\mathbf{x}^{(0)}$. Display now the position of the new iteration point on the contour plot using the option P(oint. The display shows that we have moved downhill to a point with coordinates $(0.1562, 0)$ with a function value of 0.7715.

Exercise

Use the option N(ewton line, which implements Newton's method with line searches, to find the minimum of

$$F(\mathbf{x}) = \exp(x_1^2) + \exp(0.5x_2^2)$$

Although Newton's method without this modification may generate points where the function has increased, the directions generated by it are initially downhill if $\mathbf{G}$ is positive definite. To see this, consider the change in F due to a small step s along $\mathbf{d}^{(i)}$. We have

$$F(\mathbf{x}^{(i)} + s\mathbf{d}^{(i)}) \approx F(\mathbf{x}^{(i)}) + s\mathbf{d}^{(i)t}\mathbf{g}^{(i)}$$

But, from equation 6.2.2

$$\mathbf{G}^{(i)}\mathbf{d}^{(i)} = -\mathbf{g}^{(i)}$$

so, substituting for $\mathbf{g}^{(i)}$:

$$F(\mathbf{x}^{(i)} + s\mathbf{d}^{(i)}) \approx F(\mathbf{x}^{(i)}) - s\mathbf{d}^{(i)t}\mathbf{G}^{(i)}\mathbf{d}^{(i)}$$

By the definition of the positive definiteness of $\mathbf{G}$ the second term on the right hand side is positive if s is positive, so long as $\mathbf{d}^{(i)}$ is non-zero. Hence for s sufficiently small:

$$F(\mathbf{x}^{(i)} + s\mathbf{d}^{(i)}) < F(\mathbf{x}^{(i)}) \tag{6.2.4}$$

It follows that so long as $\mathbf{G}$ is positive definite a line search along $\mathbf{d}^{(i)}$ will always be able to find a point with a lower function value than $F(\mathbf{x}^{(i)})$.

The problem remains that at a given iteration the hessian may be singular in which case equation 6.2.2 cannot be solved, or it may be indefinite, making the iterations converge to a saddle point. To overcome this problem, Levenberg and Marquardt, see for example references [6] and [7], suggested modifying the hessian by adding to it a multiple of the unit matrix. Thus the modified search direction is:

$$\mathbf{d}^{(i)} = - \left[\mathbf{G}^{(i)} + \mu \mathbf{I} \right]^{-1} \mathbf{g}^{(i)}$$

where μ is a parameter and the new iteration point is:

$$\mathbf{x}^{(i+1)} = \mathbf{x}^{(i)} + s^{(i)}\mathbf{d}^{(i)}$$

where $s^{(i)}$ is determined by finding the value of s which minimises

$$F(\mathbf{x}^{(i)} + s^{(i)}\mathbf{d}^{(i)})$$

Example

Consider

$$F(\mathbf{x}) = -x_1^4 + 2x_1^2 + x_2^2$$

The gradient is:

$$\mathbf{g} = \left[\begin{array}{c} 4x_1(1 - x_1^2) \\ 2x_2 \end{array} \right]$$

The hessian is:

$$\mathbf{G} = \left[\begin{array}{cc} 4 - 12x_1^2 & 0 \\ 0 & 2 \end{array} \right]$$

The stationary points of F are a minimum at $(0,0)$ and saddle points at $(-1,0)$, $(1,0)$.

Let us display the contours of the function in the region:

$$-0.5 \leq x_1 \leq 1.5, \qquad -0.5 \leq x_2 \leq 1.5$$

Using the program CONTOUR we will illustrate the effect on the search directions of Levenberg–Marquardt modification to the hessian matrix. At

$$\mathbf{x}^{(0)} = \begin{bmatrix} 0.8 \\ 0.5 \end{bmatrix}$$

the gradient is: $\mathbf{g}^{(0)} = \begin{bmatrix} 1.152 \\ 1 \end{bmatrix}$ and the hessian is: $\mathbf{G}^{(0)} = \begin{bmatrix} -3.68 & 0 \\ 0 & 2 \end{bmatrix}$

Thus the hessian is indefinite at this trial point; if we use it to determine the new search direction, we get:

$$\mathbf{d}^{(0)} = -\begin{bmatrix} -3.68 & 0 \\ 0 & 2 \end{bmatrix} \begin{bmatrix} 1.152 \\ 1 \end{bmatrix} = \begin{bmatrix} 0.313 \\ -0.5 \end{bmatrix}$$

Plotting $\mathbf{d}^{(0)}$ from $\mathbf{x}^{(0)}$ we find that the search direction moves the iterations towards the saddle point of F.

The Levenberg–Marquardt modification produces a new matrix $\mathbf{G}^{(0)'}$:

$$\mathbf{G}^{(0)'} = \begin{bmatrix} -3.68 & 0 \\ 0 & 2 \end{bmatrix} + \mu \begin{bmatrix} 1 & 0 \\ 0 & 1 \end{bmatrix}$$

To make $\mathbf{G}^{(0)'}$ positive definite the only relevant choices for μ are values greater than 3.68. Setting $\mu = 4$, we get:

$$\mathbf{d}^{(0)} = -\begin{bmatrix} 3.125 & 0 \\ 0 & 0.1667 \end{bmatrix} + \mu \begin{bmatrix} 1.152 \\ 1 \end{bmatrix} = -\begin{bmatrix} 3.6 \\ 0.167 \end{bmatrix}$$

Plotting this vector from the last iteration point we see that it is a good search direction.

As μ takes larger values the search directions become closer to the steepest descent direction. Try for yourself other values for μ which should illustrate that.

A succesful implementation of the method depends on efficient ways of choosing μ, which is not an easy problem.

Newton's method, and the Levenberg–Marquardt modification of it, suffer from the following disadvantages:

- second derivatives will often be unavailable or expensive to compute

- a system of linear equations must be solved on each iteration, which may be expensive for problems of large dimension

- far from a local minimum the quadratic approximation upon which Newton's method is based may be a poor one, so that much of the effort may be wasted.

Despite these disadvantages, the good ultimate convergence of Newton's method has provided the incentive for the development of the more efficient algorithms which will be described in the next section.

> 6.3   Quasi-Newton methods

We would like to avoid the calculation of the hessian matrix. One approach is to approximate the hessian by a calculation using only a knowledge of the gradient of the function at several points. For example, suppose that at the arbitrary points $\mathbf{x}^{(0)}, \mathbf{x}^{(1)}, \cdots, \mathbf{x}^{(n)}$, we know the gradients $\mathbf{g}^{(0)}, \mathbf{g}^{(1)}, \cdots, \mathbf{g}^{(n)}$ of a quadratic function $Q(\mathbf{x})$. Then in principle we could compute the hessian of Q by noting that since:

$$Q(\mathbf{x}) = a + \mathbf{b}^t\mathbf{x} + \frac{1}{2}\mathbf{x}^t\mathbf{H}\mathbf{x}$$

from which we get for the gradient vector:

$$\mathbf{g} = \mathbf{b} + \mathbf{H}\mathbf{x}$$

Therefore:

$$\mathbf{g}^{(i)} - \mathbf{g}^{(i-1)} = \mathbf{H}(\mathbf{x}^{(i)} - \mathbf{x}^{(i-1)})$$

or:

$$\gamma^{(i)} = \mathbf{H}\delta^{(i)} \qquad (6.3.1)$$

where:

$$\gamma^{(i)} = \mathbf{g}^{(i)} - \mathbf{g}^{(i-1)}$$

and

$$\delta^{(i)} = \mathbf{x}^{(i)} - \mathbf{x}^{(i-1)}$$

Thus:

$$\begin{array}{rcl}
\gamma^{(1)} & = & \mathbf{H}\delta^{(1)} \\
\gamma^{(2)} & = & \mathbf{H}\delta^{(2)} \\
\vdots & & \vdots \\
\gamma^{(n)} & = & \mathbf{H}\delta^{(n)}
\end{array} \qquad (6.3.2)$$

which can be written:

$$\mathbf{\Gamma} = \mathbf{H}\mathbf{\Delta}$$

where

$$\mathbf{\Gamma} = [\gamma^{(1)}, \gamma^{(2)}, \cdots, \gamma^{(n)}]$$

and

$$\mathbf{\Delta} = [\delta^{(1)}, \delta^{(2)}, \cdots, \delta^{(n)}]$$

Hence, we could compute $\mathbf{H}$ from:

$$\mathbf{H} = \mathbf{\Gamma}\mathbf{\Delta}^{-1} \tag{6.3.3}$$

or more usefully, assuming that $\mathbf{\Gamma}$ is non-singular:

$$\mathbf{H}^{-1} = \mathbf{\Delta}\mathbf{\Gamma}^{-1} \tag{6.3.4}$$

Then we could use Newton's method to obtain the minimum of Q in one step using:

$$\mathbf{x}^{(n+1)} = \mathbf{x}^{(n)} - \mathbf{\Delta}\mathbf{\Gamma}^{-1}\mathbf{g}^n$$

Thus we would have computed a Newton step without the explicit calculation of the hessian. However, this step can only be applied after $n + 1$ evaluations of $\mathbf{g}$ and in addition we still have to solve the system of linear equations 6.3.3, just as in Newton's method.

For these reasons, the method just described is impractical but is useful as a basic idea from which to derive procedures which make more efficient use of the gradient information and which avoid the need to solve equations.

Methods have been devised which estimate the inverse of the hessian matrix, using formulae which improve the estimate with each iteration; these are the Quasi-Newton or Variable Metric methods.

Let us first review the search directions produced by:

- steepest descent:

$$\mathbf{d}^{(i)} = -\mathbf{I}\mathbf{g}^{(i)}$$

- Newton's method:

$$\mathbf{d}^{(i)} = -(\mathbf{G}^{(i)})^{-1}\mathbf{g}^{(i)}$$

These methods generate the search directions by multiplying the gradient vector by a matrix (although in the case of steepest descent the multiplication by I is implicit). We could consider them as particular examples of a general class of methods of the form:

$$\mathbf{d}^{(i)} = -\mathbf{M}^{(i)}\mathbf{g}^{(i)}$$

The Quasi-Newton methods generate directions $\mathbf{d}^{(i)}$ using a sequence of matrices $\{\mathbf{M}^{(i)}\}$ which are made to satisfy the condition on $\mathbf{H}^{-1}$ given by equation 6.3.1, that is:

$$\mathbf{M}^{(k)}\gamma^{(k)} = \delta^{(k)} \tag{6.3.5}$$

There are various quasi-Newton methods. The original is attributed to Davidon, Fletcher and Powell, see for example reference [8], and we will call it the DFP algorithm.

> 6.3.1   The DFP algorithm

The DFP algorithm is as follows:

1. Select some arbitrary starting point $\mathbf{x}^{(0)}$ and positive definite matrix $\mathbf{M}^{(0)}$ of dimension $n \times n$.

   A common choice is
   $$\mathbf{M}^{(0)} = \mathbf{I}$$

   Set ϵ_1 and ϵ_2 as two small positive numbers.
   Calculate $\mathbf{g}^{(0)}$. Stop if
   $$||\mathbf{g}^{(0)}|| < \epsilon_1$$

   Set $k = 0$.

2. Calculate $\mathbf{p}^{(k)} = -\mathbf{M}^{(k)}\mathbf{g}^{(k)}$.

3. Compute the new point $\mathbf{x}^{(k+1)}$ by searching for the minimum along the direction $\mathbf{p}^{(k)}$.

4. Determine
   $$\delta^{(k+1)} = \mathbf{x}^{(k+1)} - \mathbf{x}^{(k)}$$

   Stop if
   $$||\delta^{(k+1)}|| < \epsilon_2$$

5. Calculate $g^{(k+1)}$. Stop if $||g^{(k+1)}|| < \epsilon_1$. Otherwise calculate

$$\gamma^{(k+1)} = g^{(k+1)} - g^{(k)}$$

6. Calculate

$$A^{(k+1)} = \frac{\delta^{(k+1)}\delta^{(k+1)^t}}{\delta^{(k+1)^t}\gamma^{(k+1)}}$$

$$B^{(k+1)} = -\frac{M^{(k)}\gamma^{(k+1)}\gamma^{(k+1)^t}M^{(k)}}{\gamma^{(k+1)^t}M^{(k)}\gamma^{(k+1)}}$$

$$M^{(k+1)} = M^{(k)} + A^{(k+1)} + B^{(k+1)}$$

7. Set $k = k + 1$ and repeat from 2.

Let us verify that the DFP formula satisfies equation 6.3.1. We have:

$$M^{(k+1)}\gamma^{(k+1)} = M^{(k)}\gamma^{(k+1)} + A^{(k+1)}\gamma^{(k+1)} + B^{(k+1)}\gamma^{(k+1)}$$

and

$$A^{(k+1)}\gamma^{(k+1)} = \frac{\delta^{(k+1)}\delta^{(k+1)^t}\gamma^{(k+1)}}{\delta^{(k+1)^t}\gamma^{(k+1)}} = \delta^{(k+1)}$$

$$B^{(k+1)}\gamma^{(k+1)} = -\frac{M^{(k)}\gamma^{(k+1)}\gamma^{(k+1)^t}M^{(k)}\gamma^{(k+1)}}{\gamma^{(k+1)^t}M^{(k)}\gamma^{(k+1)}} = -M^{(k)}\gamma^{(k+1)}$$

Thus:

$$M^{(k+1)}\gamma^{(k+1)} = M^{(k)}\gamma^{(k+1)} + \delta^{(k+1)} - M^{(k)}\gamma^{(k+1)} = \delta^{(k+1)}$$

as required.

The DFP method has the following properties:

1. Only first derivatives are needed.

2. There is no need to solve equations.

3. It can be shown that under certain assumptions, see reference [10], the matrices $M^{(i)}$ are positive definite; therefore the search directions are downhill.

4. On n-dimensional positive definite quadratic functions the search directions are conjugate. Thus on such functions at most n iterations are required.

5. On positive definite quadratic functions

$$\mathbf{M}^{(n)} = \mathbf{G}^{-1}$$

Thus DFP overcomes the disadvantages listed at the end of the previous section.

Example

Let us apply the DFP method to the function

$$F(\mathbf{x}) = \exp(x_1^2) + \exp(0.5x_2^2)$$

starting at the point

$$\mathbf{x}^{(0)} = \begin{bmatrix} 1 \\ 1 \end{bmatrix}$$

For the first iteration, we have set:

$$\mathbf{M}^{(0)} = \mathbf{I}$$

Then:

$$\mathbf{g}^{(0)} = \begin{bmatrix} 5.430 \\ 1.649 \end{bmatrix}$$

The first search is in the direction of steepest descent:

$$\mathbf{p}^{(0)} = -\mathbf{I}\mathbf{g}^{(0)} = -\mathbf{g}^{(0)}$$

The value of s which minimises

$$F(\mathbf{x}^{(0)} + s\mathbf{p}^{(0)})$$

is calculated to be 0.2065.

Thus:

$$\mathbf{x}^{(1)} = \begin{bmatrix} -0.1224 \\ 0.6596 \end{bmatrix}, \text{ and } \mathbf{g}^{(1)} = \begin{bmatrix} -0.2485 \\ 0.8199 \end{bmatrix}$$

then

$$\delta^{(1)} = \begin{bmatrix} -1.1224 \\ -0.3404 \end{bmatrix}, \text{ and } \gamma^{(1)} = \begin{bmatrix} -5.6785 \\ -0.8301 \end{bmatrix}$$

Then

$$\mathbf{A}^{(1)} = \frac{1}{6.656} \begin{bmatrix} 1.260 & 0.382 \\ 0.382 & 0.116 \end{bmatrix}$$

k	$\mathbf{x}^{(k)}$	$F^{(k)}$	$\mathbf{g}^{(k)}$	$\mathbf{M}^{(k)}$
0	$\begin{bmatrix} 1.0 \\ 1.0 \end{bmatrix}$	4.37	$\begin{bmatrix} 5.430 \\ 1.649 \end{bmatrix}$	$\begin{bmatrix} 1 & 0 \\ 0 & 1 \end{bmatrix}$
1	$\begin{bmatrix} -0.1224 \\ 0.6596 \end{bmatrix}$	2.258	$\begin{bmatrix} -0.2485 \\ 0.8199 \end{bmatrix}$	$\begin{bmatrix} 0.2099 & -0.0854 \\ -0.0854 & 0.9966 \end{bmatrix}$
2	$\begin{bmatrix} -0.02519 \\ -0.00734 \end{bmatrix}$	2.001	$\begin{bmatrix} -0.0504 \\ -0.0073 \end{bmatrix}$	$\begin{bmatrix} 0.2089 & -0.0675 \\ -0.0675 & 0.7901 \end{bmatrix}$
3	$\begin{bmatrix} 0.0001 \\ -0.0010 \end{bmatrix}$	2.000	$\begin{bmatrix} 0.0003 \\ -0.0010 \end{bmatrix}$	$\begin{bmatrix} 0.4976 & 0.0222 \\ 0.0222 & 0.8143 \end{bmatrix}$

Table 6.3.1 Iterations for the DFP method.

and

$$\mathbf{B}^{(1)} = \frac{1}{32.93} \begin{bmatrix} -32.24 & -4.714 \\ -4.714 & -0.689 \end{bmatrix}$$

which gives

$$\mathbf{M}^{(1)} = \begin{bmatrix} 1 & 0 \\ 0 & 1 \end{bmatrix} + \frac{1}{6.656} \begin{bmatrix} 1.260 & 0.382 \\ 0.382 & 0.116 \end{bmatrix} + \frac{1}{32.93} \begin{bmatrix} -32.24 & -4.714 \\ -4.714 & -0.689 \end{bmatrix}$$

that is

$$\mathbf{M}^{(1)} = \begin{bmatrix} 0.21 & -0.085 \\ -0.085 & 0.997 \end{bmatrix}$$

Table 6.3.1 displays the iterations.

Exercise

For the function:

$$F(\mathbf{x}) = \exp(x_1^2) + \exp(0.5x_2^2)$$

- Find its minimum using the option **D(fp**.

- Find its minimum using the option **C(onjug**.

- Compare the search directions given by the two methods.

> 6.3.2   The BFGS algorithm

The BFGS algorithm was developed by Broyden, Fletcher, Goldfarb and
Shanno; it is an alternative formula to the DFP method. Both methods,
when exact line searches are used, give the same sequence of points.
   The BFGS algorithm is as follows:

1. Select some arbitrary starting point $\mathbf{x}^{(0)}$ and positive definite
   matrix $\mathbf{M}^{(0)}$ of dimension $n \times n$. A common choice is

$$\mathbf{M}^{(0)} = \mathbf{I}$$

   Set ϵ_1 and ϵ_2 as two small positive numbers.
   Calculate $\mathbf{g}^{(0)}$. Stop if

$$\|\mathbf{g}^{(0)}\| < \epsilon_1$$

   Set $k = 0$.
2. Calculate $\mathbf{p}^{(k)} = -\mathbf{M}^{(k)}\mathbf{g}^{(k)}$.
3. Compute the new point $\mathbf{x}^{(k+1)}$ by searching for the minimum
   along the direction $\mathbf{p}^{(k)}$.
4. Determine
$$\delta^{(k+1)} = \mathbf{x}^{(k+1)} - \mathbf{x}^{(k)}$$
   Stop if
$$\|\delta^{(k+1)}\| < \epsilon_2$$
5. Calculate $\mathbf{g}^{(k+1)}$. Stop if $\|\mathbf{g}^{(k+1)}\| < \epsilon_1$. Otherwise calculate

$$\gamma^{(k+1)} = \mathbf{g}^{(k+1)} - \mathbf{g}^{(k)}$$

6. Calculate

$$\mathbf{A}^{(k+1)} = \left(1 + \frac{\gamma^{(k+1)^t}\mathbf{M}^{(k)}\gamma^{(k+1)}}{\delta^{(k+1)^t}\gamma^{(k+1)}}\right) \frac{\delta^{(k+1)}\delta^{(k+1)^t}}{\delta^{(k+1)^t}\gamma^{(k+1)}}$$

$$\mathbf{B}^{(k+1)} = -\frac{\delta^{(k+1)}\gamma^{(k+1)^t}\mathbf{M}^{(k)} + \mathbf{M}^{(k)}\gamma^{(k+1)}\delta^{(k+1)^t}}{\delta^{(k+1)^t}\gamma^{(k+1)}}$$

$$\mathbf{M}^{(k+1)} = \mathbf{M}^{(k)} + \mathbf{A}^{(k+1)} + \mathbf{B}^{(k+1)}$$

7. Set $k = k + 1$ and repeat from 2

Exercise

Apply the **B(fgs** option to the function

$$F(\mathbf{x}) = \exp(x_1^2) + \exp(0.5x_2^2)$$

starting at the point

$$\mathbf{x}^{(0)} = \begin{bmatrix} 1 \\ 1 \end{bmatrix}$$

and verify that the method generates the same points as those given in Table 6.3.1

> **6.3.3   Inexact line searches**

The search for efficient and reliable algorithms has led us to the DFP and BGFS methods. At this stage it is appropriate to ask to what extent this efficiency depends on accurate determination of the minimum along the line.

For example, if the quadratic search algorithm of Section 3.2.4 is used, accurate determination of the line minimum will usually require several iterations. In practice, it has been found that it may be more efficient overall to stop the line search after one or two iterations and accept the point generated as the starting point for the next search direction. This may result in more iterations of the minimisation algorithm being required but the total number of function evaluations required to find the local optimum will usually be fewer.

The reason for the increase in efficiency is that when $\mathbf{x}^{(i)}$ is far from the minimum the quality of the search direction may be poor because the function might be far from quadratic. Hence the position of the minimum on the line might not give better information on the location of the local minimum than that given by an inexact search. In addition, again because the function is far from quadratic, the search for the minimum may take many function evaluations.

On the other hand, once the quadratic approximation is applied to a point sufficiently close to a minimum it will give a good estimate of the position of this minimum. Thus few iterations of an algorithm such as the quadratic or cubic search are required to determine the position of the minimum.

There remains the problem of determining the stopping criteria for the inexact line search. To guarantee convergence to a local minimum the line search must be designed to satisfy two conditions:

- the step size must not be too small

- the reduction in the function value must not be too small.

A mathematical description of these conditions is referred to as the Wolfe–Powell conditions; see reference [10]. Most current numerical optimisation codes use inexact line searches which are designed to satisfy them.

> 6.4   The least squares problem

Up to this point we have developed techniques to determine the optimum of fairly general functions. However, it is sometimes possible to increase efficiency by taking advantage of special features which may be present in the problem. A very important class of problems for which this is possible is the minimisation of a function formed by a sum of squares of other functions. This is called the least squares problem.

The function F we seek to minimise takes the form:

$$F(\mathbf{x}) = f_1^2(\mathbf{x}) + f_2^2(\mathbf{x}) + \cdots + f_m^2(\mathbf{x}) \qquad (6.4.1)$$

where the vector $\mathbf{x}$ has n components. Such functions occur frequently in curve fitting problems. For example, to fit the function $y(t)$ which depends on the unknown parameters x_1 and x_2,

$$y = x_1 + \exp(x_2 t)$$

to a set of m data points (y_i, t_i), the values of x_1 and x_2 are chosen such that they minimise the function

$$F(\mathbf{x}) = \sum_{i=1}^{m} [y_i - (x_1 + \exp(x_2 t_i))]^2 \qquad (6.4.2)$$

Although any of the methods derived so far can be used to minimise this function, the sum of squares form can be exploited to derive an algorithm based on Newton's method which is usually more efficient.

> 6.4.1   The Gauss–Newton Method

The sum of squares function 6.4.1 can be written as:

$$F(\mathbf{x}) = \sum_{i=1}^{m} f_i^2(\mathbf{x}) \qquad (6.4.3)$$

Let us consider an iteration of Newton's method applied to the minimisation of F.

We need to calculate the gradient vector and the hessian matrix of F. Differentiating F, we get:

$$\frac{\partial F}{\partial x_1} = 2f_1\frac{\partial f_1}{\partial x_1} + 2f_2\frac{\partial f_2}{\partial x_1} + \cdots + 2f_m\frac{\partial f_m}{\partial x_1}$$

$$\vdots = \vdots$$

$$\frac{\partial F}{\partial x_n} = 2f_1\frac{\partial f_1}{\partial x_n} + 2f_2\frac{\partial f_2}{\partial x_n} + \cdots + 2f_m\frac{\partial f_m}{\partial x_n}$$

Now we find it useful to define a new vector $\mathbf{f}$ as

$$\mathbf{f} = \begin{bmatrix} f_1 \\ f_2 \\ \vdots \\ f_m \end{bmatrix}$$

Then we can write:

$$F = \mathbf{f}^t\mathbf{f}$$

and

$$\begin{bmatrix} \frac{\partial F}{\partial x_1} \\ \vdots \\ \frac{\partial F}{\partial x_n} \end{bmatrix} = 2\begin{bmatrix} \frac{\partial f_1}{\partial x_1} & \frac{\partial f_2}{\partial x_1} & \cdots & \frac{\partial f_m}{\partial x_1} \\ \vdots & \vdots & & \vdots \\ \frac{\partial f_1}{\partial x_n} & \frac{\partial f_2}{\partial x_n} & \cdots & \frac{\partial f_m}{\partial x_n} \end{bmatrix}\begin{bmatrix} f_1 \\ f_2 \\ \vdots \\ f_m \end{bmatrix} = 2\mathbf{J}^t\mathbf{f}$$

where $\mathbf{J}$ is the matrix with components $J_{ij} = \partial f_i/\partial x_j$ and is called the *Jacobian matrix* of F.

Differentiating again:

$$\frac{\partial^2 F}{\partial x_1^2} = 2\left[\left(\frac{\partial f_1}{\partial x_1}\right)^2 + f_1\frac{\partial^2 f_1}{\partial x_1^2}\right] + 2\left[\left(\frac{\partial f_2}{\partial x_1}\right)^2 + f_2\frac{\partial^2 f_2}{\partial x_1^2}\right] + \cdots$$

$$+ 2\left[\left(\frac{\partial f_m}{\partial x_1}\right)^2 + f_m\frac{\partial^2 f_m}{\partial x_1^2}\right]$$

$$\frac{\partial^2 F}{\partial x_2\partial x_1} = 2\left[\frac{\partial f_1}{\partial x_2}\frac{\partial f_1}{\partial x_1} + f_1\frac{\partial^2 f_1}{\partial x_2\partial x_1}\right] + 2\left[\frac{\partial f_2}{\partial x_2}\frac{\partial f_2}{\partial x_1} + f_2\frac{\partial^2 f_2}{\partial x_2\partial x_1}\right] + \cdots$$

$$+ 2\left[\frac{\partial f_m}{\partial x_2}\frac{\partial f_m}{\partial x_1} + f_m\frac{\partial^2 f_m}{\partial x_2\partial x_1}\right]$$

and similarly for the other elements of the hessian matrix.

By separating into two groups the terms with first derivatives and the terms with second derivatives, we can write the hessian matrix **G** as follows:

$$\mathbf{G} = 2\left[\mathbf{J}^t\mathbf{J} + \sum_{i=1}^{m} f_i \mathbf{G}_i\right] \tag{6.4.4}$$

where $\mathbf{G}_i$ is the hessian matrix of the function f_i.

The Gauss–Newton method is based on the assumption that the second term in equation 6.4.4 is negligible in comparison with the first.

This assumption is often justifiable on two grounds:

- the absolute values of the f_i are expected to be small in the vicinity of the minimum of F

- in many practical cases the second derivatives of the functions f_i are small.

These arguments lend support to the Gauss–Newton assumption which cannot, however, be justified solely on mathematical grounds. In fact the best support for the assumption is that methods based upon it are found to perform well in practice.

The Gauss–Newton algorithm is obtained by replacing the hessian by $2\mathbf{J}^t\mathbf{J}$ in Newton's method. As with Newton's method we modify the algorithm such that the step length is selected using a line search.

The modified Gauss–Newton method is as follows:

1. Select some arbitrary starting point $\mathbf{x}^{(0)}$. Set $k = 0$.

2. Calculate
$$\mathbf{g}^{(k)} = 2\mathbf{J}^{(k)^t}\mathbf{f}^{(k)}.$$
Stop if $|\mathbf{g}^{(k)}| < \epsilon_1$. Otherwise calculate $\mathbf{J}^{(k)^t}\mathbf{J}^{(k)}$.

3. Solve for $\mathbf{d}^{(k)}$ the system of equations:
$$\mathbf{J}^{(k)^t}\mathbf{J}^{(k)}\mathbf{d}^{(k)} = -\mathbf{J}^{(k)^t}\mathbf{f}^{(k)}$$

Stop if
$$|\mathbf{d}^{(k)}| < \epsilon_2$$

4. Determine
$$\mathbf{x}^{(k+1)} = \mathbf{x}^{(k)} + s\mathbf{d}^{(k)}$$
where s is ascertained by a line search, which need not be exact. Set $k = k + 1$ and repeat from 2.

The values of ϵ_1 and ϵ_2 are selected as discussed in Chapter 5. The advantages of the Gauss–Newton over Newton's algorithm are:

- No second derivatives need to be computed.

- We show below that $\mathbf{J}^t\mathbf{J}$ is inherently positive semi-definite and this, as discussed in Section 5.2, implies that search directions are never uphill.

To prove that the matrix $\mathbf{J}^t\mathbf{J}$ is positive semi-definite we require to show that
$$\mathbf{z}^t\mathbf{J}^t\mathbf{J}\mathbf{z} \geq 0$$
for all non-zero vectors $\mathbf{z}$. By setting

$$\mathbf{y} = \mathbf{J}\mathbf{z}$$

and substituting in the previous equation, we get

$$\mathbf{z}^t\mathbf{J}^t\mathbf{J}\mathbf{z} = \mathbf{y}^t\mathbf{y} = \sum y_i^2 \geq 0$$

Note that when $\mathbf{J}^t\mathbf{J}$ is singular then the vector $\mathbf{y}$ may be zero even though $\mathbf{z}$ is non-zero.

Exercise

Use the program LSQ with the following set of data points to determine the values of x_1 and x_2 such that the model:

$$y = x_1 + \exp\left(x_2 t\right)$$

minimises the function given in equation 6.4.2.

t	0.0	0.1	0.25	0.3	0.42
y	2.4724	1.4861	1.3555	1.7444	1.21

Table 6.4.1 Data points.

> 6.4.2   Singularity in the J^tJ matrix

The applicability of the method depends on the relative values of m and n. The algorithm is designed for use when $m > n$. For $m = n$ the algorithm takes a simplified form which is discussed in the next section. When $m < n$ the matrix J^tJ is inherently singular, even though the true hessian might not be. In this case therefore the method cannot be applied.

However even when $m > n$ there still remains the possibility that the matrix J^tJ is singular or near singular at some trial point $x^{(k)}$. The Levenberg–Marquardt modification, which we discussed on page 193, is particularly useful for this case because *any* positive value of the parameter μ will suffice to make the modified matrix positive definite. The reason for this is that as J^tJ is a positive semi-definite matrix it has non-negative eigenvalues, and so adding a positive quantity however small to each element on the diagonal will cause the modified matrix to have all eigenvalues positive.

At a given iteration, the need for the modification may become apparent during the process of solving the system of linear equations which determine the search direction. When the equations happen to be singular and when for example gaussian elimination is used to solve them e.g. [2], at same stage in the calculations, the diagonal of the transformed matrix will have zero values. At this stage the solution process is abandoned and the Levenberg–Marquard modification is introduced. A value of μ can then be computed as:

$$\mu = 10^{-4} \sum_{i=1}^{n} (J^tJ)_{ii}$$

i.e., a small fraction of the sum of the diagonal elements of the J^tJ matrix. The factor 10^{-4} has been found in practice to be suitable. The modified equations can now be solved to determine the new search direction.

The modified Gauss–Newton method is nearly always the best method to use for least squares problems. However, it is important to be aware that the method may not converge in certain circumstances. When the second term in equation 6.4.4 is not negligible in the vicinity of a local optimum the sequence of points generated may spiral indefinitely about the local minimum. This may occur whenever the values of the subfunctions f_i are large at the minimum of F or whenever the f_i have hessian matrices with large elements. In such cases we should use some of the other methods discussed in previous chapters.

Exercise

Starting at $(-1, 1.1)$ find the minimum of Rosenbrock's function using:

- the option **B(fgs** in the program NEWTON

- the option **G(aunew** in the program LSQ.

    Recall that to use this option you need to enter Rosenbrock's function as a sum of squares using the program PREPARE.

Note how much more effective the Gauss–Newton method is. This will normally be the case.

> 6.4.3   Solving nonlinear equations by least squares

Let us consider the following problem:
Find **x** such that

$$f_1(\mathbf{x}) = 0$$
$$f_2(\mathbf{x}) = 0$$
$$\vdots \quad \vdots$$
$$f_n(\mathbf{x}) = 0$$

where $f_i(\mathbf{x})$ are nonlinear functions. This is the n-dimensional nonlinear equation problem and can be solved by minimising the following sum of squares function:

$$F(\mathbf{x}) = f_1^2(\mathbf{x}) + f_2^2(\mathbf{x}) + \cdots + f_n^2(\mathbf{x})$$

This function corresponds to a least squares problem where $m = n$ and it leads to a simplification of the Gauss–Newton algorithm.

In this case the matrix **J** is square, so the system of linear equations to be solved for the search direction can be simplified from

$$\mathbf{J}^{(k)^t}\mathbf{J}^{(k)}\mathbf{d}^{(k)} = -\mathbf{J}^{(k)^t}\mathbf{f}$$

to

$$\mathbf{J}^{(k)}\mathbf{d}^{(k)} = -\mathbf{f}$$

by multiplying each side by $(\mathbf{J}^{(k)^t})^{-1}$.

In this form the Gauss–Newton algorithm is known as the Newton–Raphson method.

> 6.5   Summary

In this chapter we have discussed methods which make use of second derivative information. All of these methods are based on the Newton algorithm; the Quasi-Newton methods have been developed in order to overcome the disadvantage in this basic algorithm. The Gauss–Newton algorithm does this in a different way for the least squares problem by exploiting the special form of the problem.

When evaluating these methods using the programs provided with this volume, it must be remembered that only two-dimensional examples can be solved. The algorithms can of course be applied to problems in many variables. In such cases the differences in performance among the various algorithms may be more marked than in the examples given.

Although there is no theoretical upper limit on the dimensions of the problems which can be solved, in practice it is usual to regard these algorithms as suitable for problems with fewer than 100 variables.

> Chapter 7

> Optimisation in practice

> 7.1 Introduction

We have introduced the basic concepts of numerical optimisation and have described techniques for finding the solution of optimisation problems which do not involve constraints. These techniques are guaranteed to find local minima of 'well behaved' functions. However when we come to apply such techniques to a wide range of practical problems we find that there are additional considerations which influence the efficiency of the methods and the usefulness of the results. In this chapter we will consider some of these.

The first problem arises from the fact that numerical optimisation is usually carried out using digital computers, which can represent and store numbers with only a limited degree of precision. This in turn limits the accuracy with which the position of a local minimum can be located, and in the first section of this chapter we shall discuss some of the implications of this fact.

A second practical consideration concerns the use of estimated first derivatives of the function to be minimised, when analytical expressions are not available. Such estimates can be made using finite differences of the function, which in effect means using additional function evaluations to make up for the lack of exact expressions for the derivatives. The questions which arise concern the adequacy of such estimates, and how the efficiency of gradient methods using estimated derivatives compares with that of direct search methods, which avoid the calculation of derivatives altogether.

Finally we turn to one of the fundamental limitations of all the meth-

105

ods we have introduced, namely that they are designed to locate only local optima. While this is often acceptable, there are many cases in which the global optimum is required. When a function has several local minima, the one which is found by a local minimisation algorithm depends on the starting point used; we shall see how this property can be used to design algorithms which have an increased likelihood of locating the global optimum.

Finally, we shall discuss how to select a suitable method for solving any given problem.

> 7.2   Accuracy of the solution

Let us first consider the effects of errors in evaluating F on the determination of the minimum. Suppose it is known that, for some reason, F can only be computed with limited accuracy, so that at the solution $\mathbf{x}^{(*)}$ the computed value of F is

$$F_{comp}(\mathbf{x}^{(*)}) = F(\mathbf{x}^{(*)}) + \epsilon$$

where ϵ is the unknown error. Then any point $\bar{\mathbf{x}}$ for which

$$|F(\bar{\mathbf{x}}) - F(\mathbf{x}^{(*)})| \leq |\epsilon| \tag{7.2.1}$$

must be regarded as a point which is an acceptable estimate of $\mathbf{x}^{(*)}$. Let us now determine the region which contains all such acceptable values. For this we use the Taylor's expansion with

$$\bar{\mathbf{x}} = \mathbf{x}^{(*)} + h\mathbf{e}$$

where $\mathbf{e}$ is a unit vector along any direction. We get for sufficiently small h:

$$F(\mathbf{x}^{(*)} + h\mathbf{e}) \approx F(\mathbf{x}^{(*)}) + h\mathbf{e}^{(t)}\mathbf{g}^{(*)} + \frac{1}{2}h^2\mathbf{e}^{(t)}\mathbf{G}^{(*)}\mathbf{e}$$

At the minimum the gradient is zero, so for an acceptable solution we have:

$$|F(\bar{\mathbf{x}}^{(*)}) - F(\mathbf{x}^{(*)})| \approx |\frac{1}{2}h^2\mathbf{e}^{(t)}\mathbf{G}^{(*)}\mathbf{e}| < |\epsilon|$$

Therefore

$$h^2 = ||\bar{\mathbf{x}} - \mathbf{x}^{(*)}||^2 < \frac{2|\epsilon|}{\mathbf{e}^{(t)}\mathbf{G}^{(*)}\mathbf{e}} \tag{7.2.2}$$

or

$$||\bar{\mathbf{x}} - \mathbf{x}^{(*)}|| < \frac{\sqrt{2|\epsilon|}}{(\mathbf{e}^{(t)}\mathbf{G}^{(*)}\mathbf{e})^{1/2}}$$

(Note that $G^{(*)}$ must be positive definite, so the denominator of the right hand side is positive).

Let us use an example to investigate the implications of equation 7.2.2. Consider the following simple quadratic function which has a minimum at the point $(0,0)$:

$$F(\mathbf{x}) = ax_1^2 + bx_2^2 = \frac{1}{2}\mathbf{x}^{(t)} \begin{bmatrix} 2a & 0 \\ 0 & 2b \end{bmatrix} \mathbf{x}$$

where a and b are positive constants. In this case equation 7.2.2 takes the form:

$$a\bar{x}_1^2 + b\bar{x}_2^2 < \epsilon$$

which defines the points inside the ellipse with semi-axes $\sqrt{\epsilon/a}$ and $\sqrt{\epsilon/b}$ along the coordinate axes and centred at the minimum. The worst possible error in the determination of the minimum corresponds to the points at the ends of the larger semi-axis.

In the general two-dimensional case the matrix G is not necessarily diagonal; however we have seen in Chapter 5 that positive definite quadratic functions have elliptical countours, so the maximum errors always occur at the ends of the larger semi-axis. It can be shown that the lengths of the axes are in fact proportional to the square roots of the reciprocals of the eigenvalues of G and the directions of the axes are given by the corresponding eigenvectors, see for example reference [15]

The error in F can arise from many sources, not only from the direct effect of finite precision on the representation of its value but also as a result of the various errors which might arise during the course of its calculation. In practice F may well not be representable as a simple function of $\mathbf{x}$, but may need to be computed by a lengthy numerical process. If this includes such procedures as numerical integration, solution of linear equations or interpolation then the cumulative effect of these errors on the value of F may be very great. In addition to reducing the accuracy with which the minimum may be located, such errors may on occasion even prevent the convergence of the algorithm to any point. Some of the effects of random errors in F will be discussed in the next section.

This discussion is not intended as a full treatment of the error analysis problem for numerical optimisation, which is well beyond the scope of this book. However, it should serve to stress the importance of careful programming of the function evaluation routine when using a numerical optimisation program, and to provide a warning against uncritical acceptance of the results of any numerical computation.

> 7.3 Estimation of first derivatives

When formulae for the first derivatives of F are not available these can be estimated using finite differences. Two forms are possible:

- Forward differences. The estimated derivative is

$$\frac{\partial \hat{F}}{\partial x_i} = \frac{F(\mathbf{x} + h\mathbf{e}_i) - F(\mathbf{x})}{h}$$

where $\mathbf{e}_i$ is a unit vector along the i-th axis and h is a small step.

- Central differences. The estimated derivative is

$$\frac{\partial \hat{F}}{\partial x_i} = \frac{F(\mathbf{x} + h\mathbf{e}_i) - F(\mathbf{x} - h\mathbf{e}_i)}{2h}$$

The truncation errors (that is, the errors caused by ignoring the nonlinearity of F) for the two forms are derived as follows:

$$F(\mathbf{x} + h\mathbf{e}_i) \approx F(\mathbf{x}) + h\frac{\partial F}{\partial x_i} + \frac{1}{2}h^2\frac{\partial^2 F}{\partial x_i^2} + \frac{1}{6}h^3\frac{\partial^3 F}{\partial x_i^3} \qquad (7.3.1)$$

$$F(\mathbf{x} - h\mathbf{e}_i) \approx F(\mathbf{x}) - h\frac{\partial F}{\partial x_i} + \frac{1}{2}h^2\frac{\partial^2 F}{\partial x_i^2} - \frac{1}{6}h^3\frac{\partial^3 F}{\partial x_i^3} \qquad (7.3.2)$$

For forward differences we subtract $F(\mathbf{x})$ from both sides of equation 7.3.1 and divide by h to give:

$$\frac{\partial \hat{F}}{\partial x_i} \approx \frac{\partial F}{\partial x_i} + \frac{h}{2}\frac{\partial^2 F}{\partial x_i^2} + \frac{1}{6}h^2\frac{\partial^3 F}{\partial x_i^3}$$

For central differences we subtract equation 7.3.2 from equation 7.3.1 and divide by $2h$ to give:

$$\frac{\partial \hat{F}}{\partial x_i} \approx \frac{\partial F}{\partial x_i} + \frac{h^2}{6}\frac{\partial^3 F}{\partial x_i^3}$$

Thus the error in forward differences is of the order of h while in central differences it is of the order of h^2 and therefore much smaller.

Forward differences require $n + 1$ function evaluations to compute all n first derivatives, while central differences require $2n$ evaluations. The reduced truncation error is therefore gained at the cost of the extra function evaluations. In practice it turns out that this cost is worth paying.

Let us now consider the choice of a value for h. We have shown that large values of h lead to large truncation errors. However, small values of h can cause rounding errors. These arise from the fact that x_i and F are represented in a digital computer by numbers of finite length. Suppose that they are represented by s significant figures. Then a change will only be significant if it affects at least the s-th significant figure. Thus we must have

$$h \geq 10^{-s} x_i$$

in order for the change in x_i to be significant.

The change in F due to a change h in x_i is given by

$$\frac{\partial F}{\partial x_i} h$$

which must be at least equal to $10^{-s}|F|$. Therefore:

$$|h| \geq 10^{-s} |F/(\frac{\partial F}{\partial x_i})|$$

In addition, as discussed in Section 7.2, the value of F might be affected by rounding errors occurring during intermediate calculations. These will in practice often be much larger than the direct effect mentioned above.

The choice of h thus involves a compromise between truncation and rounding errors. In practice most optimisation algorithms are not sensitive to the value of h. Since computers usually allow a single precision significance of 7 or 8 decimal digits, a value of h giving a change in x_i of between 0.001 and 0.1 % will be more than enough to avoid the direct effect of rounding errors, and normally will not produce unacceptably large truncation errors.

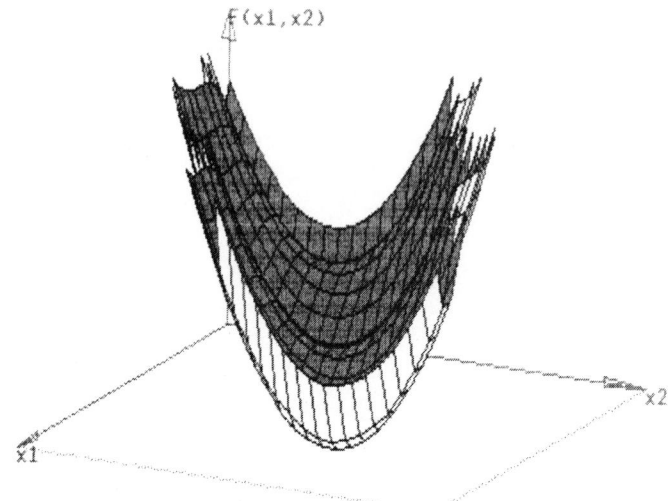

Figure 7.4.1 The three-humped camel-back function.

> 7.4 Global optimisation

Let us consider the problem of finding the global optimum of the function:

$$F(\mathbf{x}) = 2x_1^2 - 1.05x_1^4 + \frac{1}{6}x_1^6 - x_1x_2 + x_2^2$$

which is called the three-humped camel-back function [11]. We shall consider this function in the range:

$$-2 \leq x_1 \leq 2 \quad -2 \leq x_2 \leq 2$$

Figure 7.4.1 shows a perspective view of the function and Figure 7.4.2 its contours. You can use the programs CONTOUR and 3D-PLOT to reproduce these pictures.

From the pictures it is apparent that the function has three local minima in the given region. Now to find the position of these optima you should run any of the optimisation programs three times to find the position of the minimum of F, starting each time at a different one of the three following points:

$$A = (-1.5, 0) \quad B = (0.5, -0.5) \quad C = (1.5, 1.5)$$

You should discover minima at:

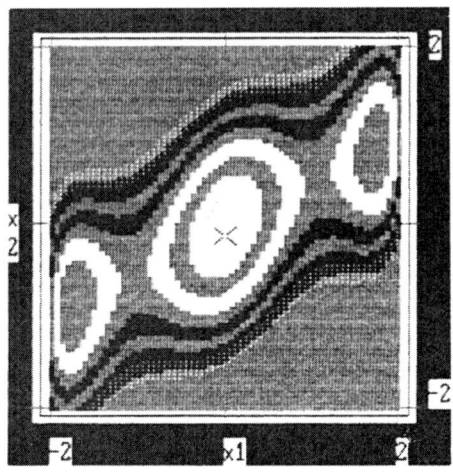

Figure 7.4.2 Contour map of the three-humped camel-back function.

Run i	Starting point	Local minimum at $x^{(*i)}$	$F(x^{(*i)})$
1	A	$(-1.748, -0.874)$	0.299
2	B	$(0,0)$	0.0
3	C	$(1.748, 0.874)$	0.299

We happen to know that this function has only three minima, so the global minimum is at $x^{(*2)}$.

In this case, where we know how many local minima there are, we were able to locate all of them by starting at different points and thus to find the global minimum.

In the general case, when dealing with more complex functions or with functions of more than two variables, it is not possible to find all the local minima in a given region by using plots of the function. The task is therefore to carry out a more or less exhaustive search for local minima and, equally importantly, to decide when to stop.

Thus it is possible to characterise the global optimisation problem in terms of the following tasks:

- how to find suitable starting points such that all local optima in the search region are found, but, if possible, none is found more than once;

- how to recognise when all the local optima have been found.

There is in fact no method which is guaranteed to find the global optimum of a general function. The difficulty lies in recognising it, because to test the assertion that a given point is the global minimum requires the evaluation of the function at every point, which is impossible. However with some restrictions to the problem useful results can still be obtained without excessive use of computing resources.

Firstly, we confine the search to a specified region simply by applying upper and lower bounds to the variables.

Secondly, we accept as a solution that point which we can regard with some specified degree of confidence as the global optimum. To implement this we need to be able to calculate the confidence that a given local optimum is also the global optimum in a given region.

Within this framework a number of global optimisation methods have been developed. Although the theory they use is too elaborate to be discussed here, the methods themselves are often simple to apply.

They fall into two classes:

- deterministic methods

- stochastic methods .

The deterministic methods seek to use some known property of the function to locate the global minimum with certainty. For example, if an upper limit can be placed on the rate at which the function value changes, this information can be used to determine the spacing of points on a grid. If the minimum function value over these grid points is then found, the position of the global minimum will be known, because the possibility of a much lower value occurring elsewhere between grid points can be eliminated. Unfortunately, the required information about the function is hardly ever available, and even if it were the grid search itself would usually be extremely time consuming. Until it is completed, however, little useful information will have been obtained. These characteristics greatly restrict the usefulness of deterministic methods.

Stochastic methods on the other hand, which treat the problem in probabilistic terms, can be applied to most problems, and the confidence which they generate in the location of the global minimum tends to increase monotonically with the effort expended. More details can be found in reference [12] .

We present one of the stochastic methods below.

> 7.4.1   The multi-level single linkage method

This method, due to Rinnooy Kan and Timmer [12], searches for the global minimum of a function $F(\mathbf{x})$ of n variables in a given search region S.
The k-th iteration of the algorithm is as follows:

1. Generate N sample points at random such that every point in the region has an equal probability of being selected.
   Evaluate F at each sample point.

2. Select starting points for local searches.
   Each sample point is selected as a starting point if it is not within a certain critical distance of a point which was previously selected and which has a lower or equal function value.
   The critical distance is given by:

$$r(k) = \pi^{-1/2} \left[\sigma m(S)\Gamma(1 + \frac{n}{2}) \frac{\ln(kN)}{kN} \right]^{\frac{1}{n}}$$

where
$$m(S) \equiv \text{ volume of S}$$
$$\sigma \equiv \text{ a positive parameter}$$
$$\Gamma \equiv \text{ the gamma function.}$$

3. Perform local minimisations from all starting points.

4. Decide whether to stop. If confidence is sufficiently high that all the local minima have been found, regard the lowest local minimum as global, otherwise repeat the iteration.

The method uses two formulae to measure confidence:

$E(T) \equiv$ the expected total number of local minima.

$E(C) \equiv$ expected fraction of S covered by the regions of attraction of the local minima found so far.

The region of attraction of a local minimum $\mathbf{x}^{(*)}$ for a local optimisation algorithm is the set of starting points from which that algorithm will converge to $\mathbf{x}^{(*)}$.

The expressions for these quantities are:

$$E(T) = \frac{w(s-1)}{s-w-2}$$

and

$$E(C) = \frac{(s-w-1)(s+w)}{s(s-1)}$$

where w is the number of distinct local minima found and s is the number of actual optimisations performed.

The stopping rule used is that the algorithm terminates after the k-th iteration if:

$$E(T) < w + 0.5$$

and

$$E(C) \geq 0.995$$

Exercise

Determine the global optimum of the three-humped camel-back function using the program GLOBAL which implements the above algorithm.

> 7.5   Selecting an optimisation method

All the numerical optimisation methods we have discussed are based on the hill-climbing analogy. To make this more realistic, consider the situation of a climber on a mountain, trying to make a descent without a map and in dense fog. He can rely on an altimeter to measure altitude, a compass, and a device to tell him how far he has moved along any direction. The altimeter takes a long time to set up and use, but movement itself is easy. The climber wishes to descend as fast as possible. What is his best strategy?

He has two main options, depending on whether he wishes to take account of the local slope of the mountain. If not, he must use a strategy corresponding to a direct search method, which consists in moving along a series of directions regularly measuring altitude until he finds an increase. Having decided how to choose his directions, the only additional decision he needs to take is how accurately to attempt to locate the lowest points along these directions. He could spend time searching backwards and forwards between the last three measurement points to locate the lowest point along that direction or alternatively he could set off along a new direction without spending time locating exactly the lowest point along the line. The latter strategy is generally best. The directions that the climber chooses might correspond to any of the ones specified by the methods of Chapter 3, such as Univariate search or DSC. It is left as an exercise for the reader to determine how this might be accomplished with the instruments available.

The second option is to use a measurement of slope to determine the direction of movement. Using his altimeter, the climber could measure the difference in altitude between points close to one another and estimate the slope along the line joining them using the methods of the previous section. With this measure of the slope he can calculate a direction in which to move using one of the gradient methods. Of course, any errors in the estimates of the slope will make his search less efficient than it would otherwise be.

Let us now follow the analogy further and assume that he is additionally equipped with a spirit level. (This is analogous, in optimisation, to making available exact expressions for the first derivatives). He can now measure the slope of the ground more precisely, with a corresponding improvement in the efficiency of the search. However, if the ground is stoney or broken his measurements may be misleading, and he may be better off avoiding the use of slope measurements altogether. (A hill with a stoney surface would correspond to a mathematical function with small random errors caused, for example, by numerical inaccuracies occurring during the evaluation process. Such functions are often referred to as noisy functions).

The choices facing the climber are similar to those experienced by an optimiser when selecting an algorithm. A simple guide for selecting a suitable method is given in Figure 7.5.1.

The figure is mostly self-explanatory. Although the choice of Gauss–Newton for sums of squares problems is fairly automatic, the figure allows for the possibility that this may not be best because of the inadequacy of the Gauss–Newton approximation in certain cases (see Chapter 6). In such cases, it is simplest to ignore the special form of the function. The Conjugate Gradient method (Chapter 4) is less efficient than the Quasi-Newton algorithms, but has the advantage of requiring very little storage space. It is therefore a possible choice when solving problems which have many variables. In fact, the technique is now often used in the solution of very large systems of linear equations by minimising residual errors because it requires no storage of matrices, and because it often produces acceptable solutions in a very small number of iterations

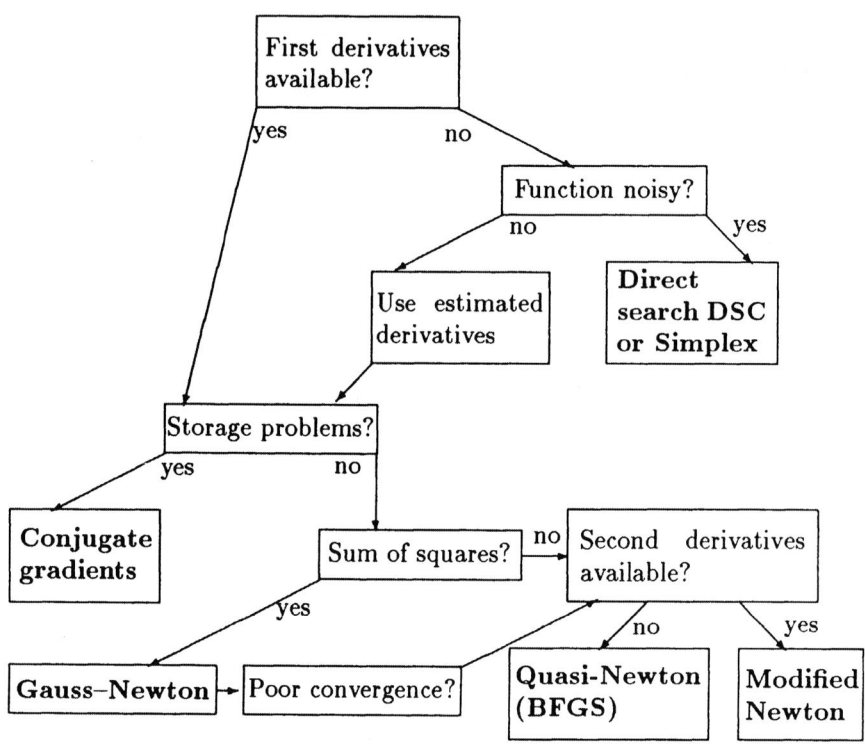

Figure 7.5.1 Guide for selecting an optimisation method.

(Jennings [5]).

> **7.6   Summary**

In this book we have introduced some of the fundamental ideas behind modern methods for solving numerical optimisation problems. At this point the reader should be able to recognise situations in which such methods might be of practical use to him, to formulate his problem accordingly and to select a suitable algorithm. Computer implementations of the algorithms are readily available in subroutine libraries such as the NAG Subroutine Library, see reference [13] and the OPTIMA library, see reference [14]. However, it must be said that the successful use of the numerical optimisation techniques does sometimes require a degree of skill which can only be gained by experience. Some of the practical decisions which must be made have been touched upon in this chapter; however, it is not possible to discuss here all the considerations which may be relevant. For example, we have not made specific mention of the need to scale the optimisation variables. In some cases failure to do this can result in a poorly conditioned problem, leading to slow convergence and an inaccurate solution. However, in practice this can usually be avoided simply by a sensible choice of units for the optimisation variables, so that the function is approximately equally sensitive with respect to each of them. Again, most computer implementations will require the user to choose the values of various parameters such as convergence tolerences, step lengths for estimated derivatives and so on. Some advice on these matters has been given at appropriate points in the book; the best approach is to be prepared to experiment with various values until a completely dependable result has been achieved. Another possible source of difficulty stems from the nature of the functions which may have to be minimised in practical situations. We have made the assumption that the function to be mimimised is "well behaved"; but in practice it may be impossible to guarantee this. In such cases the only possible approach is once again to regard the problem as an experimental one, and to be prepared to try various starting points and parameter values until consistent results are obtained. It is advisable to carry out a sensitivity analysis on the solution of any optimisation problem, at least by perturbing the values of the optimisation variables by small amounts and noting the effects on function and gradients. Ideally, the eigenvalues and eigenvectors of the objective function should be computed. If the ratio of maximum to minimum eigenvalue is very different from unity it is a measure of ill-conditioning, and the accuracy of the solution may be in doubt (see Section 7.3) (e.g. McKeown [15]).

On the positive side, it is important not to underestimate the power of numerical optimisation algorithms. In particular, it should be realised that it is not necessary to be able to express the function mathematically in order to minimise it. All that is necessary is that a subroutine can be written which computes its value for any given values of the optimisation variables. This opens the door to the optimisation of very complex systems. In practice many of these applications will involve constraints on the choice of values for the optimisation variables, and therefore they cannot make direct use of the methods described here. But many do not, particularly least squares modelling problems. Although constrained optimisation is beyond the scope of this book, it involves many of the ideas introduced here and some of the methods which we have described also form the basis for constrained methods.

Bibliography

[1] Kiefer, J. 'Sequential minimax search for a maximum', *Proc. Am. Math. Soc.*, 4 (1953).

[2] Harding, R.D. and Quinney, D.A. 'A simple introduction to numerical analysis', *Adam Hilger*, Bristol, (1986).

[3] Rosenbrock, H.H. 'An automatic method for finding the greatest or least value of a function', *The Computer Journal*, 3 (1960).

[4] Hooke, R. and Jeeves, T.A. 'Direct search solution of numerical and statistical problems', *J. Assoc. Comput. Mach.*, 8 (1961).

[5] Jennings, A. 'Matrix Computation for Engineers and Scientists', *John Wiley & Sons*, New York, (1977).

[6] Levenberg, K. 'A method for the solution of certain nonlinear problems in least squares', *Quart. Appl. Math.*, 2 (1944).

[7] Goldfeld, S.M., Quandt, R.E. and Trotter, H.F. 'Maximization by quadratic hill-climbing', *Econometrica*, 34 (1966).

[8] Fletcher, R. and Powell, M.J.D. 'Rapidly convergent descent method for minimisation', *Computer Journal*, 7 (1963).

[9] Bunday, B.D. 'Basic optimisation methods', *Edward Arnold*, London, (1984).

[10] Fletcher R. 'Practical methods of optimization', 2nd Edn *John Wiley & Sons*, New York, (1987).

[11] L.C.W. Dixon and G.P.Szegö (Eds) 'Towards global optimization', *North-Holland*, Amsterdam, (1975)

[12] Rinnooy Kan, A.H.G. and Timmer, G.T. 'Towards global optimization methods (I and II)', *Math. Prog.*, 39 (1987).

[13] The Numerical Algorithms Group Ltd, Oxford, UK.

[14] OPTIMA manual. Numerical Optimization Centre, Hatfield Polytechnic, UK.

[15] McKeown, J.J. 'Sensitivity analysis with respect to independent variables', in *Nonlinear optimisation theory and algorithms*, ed. L.C.W. Dixon, E. Spedicato and G.P.Szegö, *Birkhäser*, Stuttgart, (1980).

Index